The Captains of Energy

Systems Dynamics from an Energy Perspective

Synthesis Lectures on Engineering

Each book in the series is written by a well known expert in the field. Most titles cover subjects such as professional development, education, and study skills, as well as basic introductory undergraduate material and other topics appropriate for a broader and less technical audience. In addition, the series includes several titles written on very specific topics not covered elsewhere in the Synthesis Digital Library.

A Little Book on Teaching: A Beginner's Guide for Educators of Engineering and Applied Science
Steven F. Barrett
2012

Engineering Thermodynamics and 21st Century Energy Problems: A Textbook Companion for Student Engagement
Donna Riley
2011

MATLAB for Engineering and the Life Sciences
Joseph V. Tranquillo
2011

Systems Engineering: Building Successful Systems
Howard Eisner
2011

Fin Shape Thermal Optimization Using Bejan's Constructal Theory
Giulio Lorenzini, Simone Moretti, and Alessandra Conti
2011

Geometric Programming for Design and Cost Optimization (with illustrative case study problems and solutions), Second Edition
Robert C. Creese
2010

Survive and Thrive: A Guide for Untenured Faculty
Wendy C. Crone
2010

Geometric Programming for Design and Cost Optimization (with Illustrative Case Study Problems and Solutions)
Robert C. Creese
2009

Style and Ethics of Communication in Science and Engineering
Jay D. Humphrey and Jeffrey W. Holmes
2008

Introduction to Engineering: A Starter's Guide with Hands-On Analog Multimedia Explorations
Lina J. Karam and Naji Mounsef
2008

Introduction to Engineering: A Starter's Guide with Hands-On Digital Multimedia and Robotics Explorations
Lina J. Karam and Naji Mounsef
2008

CAD/CAM of Sculptured Surfaces on Multi-Axis NC Machine: The DG/K-Based Approach
Stephen P. Radzevich
2008

Tensor Properties of Solids, Part Two: Transport Properties of Solids
Richard F. Tinder
2007

Tensor Properties of Solids, Part One: Equilibrium Tensor Properties of Solids
Richard F. Tinder
2007

Essentials of Applied Mathematics for Scientists and Engineers
Robert G. Watts
2007

Project Management for Engineering Design
Charles Lessard and Joseph Lessard
2007

Relativistic Flight Mechanics and Space Travel
Richard F. Tinder
2006

The Captains of Energy: Systems Dynamics from an Energy Perspective
Vincent C. Prantil and Timothy Decker
www.morganclaypool.com

ISBN: 9781627055888   paperback
ISBN: 9781627055895   ebook

DOI 10.2200/S00610ED1V01Y201410ENG024

A Publication in the Morgan & Claypool Publishers series
SYNTHESIS LECTURES ON ENGINEERING

Lecture #24
Series ISSN
Print 1939-5221   Electronic 1939-523X

The Captains of Energy

Systems Dynamics from an Energy Perspective

Vincent C. Prantil
Milwaukee School of Engineering

Timothy Decker
Milwaukee Area Technical College
University of Wisconsin Milwaukee

SYNTHESIS LECTURES ON ENGINEERING #24

MORGAN & CLAYPOOL PUBLISHERS

ABSTRACT

In teaching an introduction to transport or systems dynamics modeling at the undergraduate level, it is possible to lose pedagogical traction in a sea of abstract mathematics. What the mathematical modeling of time-dependent system behavior offers is a venue in which students can be taught that physical analogies exist between what they likely perceive as distinct areas of study in the physical sciences. We introduce a storyline whose characters are superheroes that store and dissipate energy in dynamic systems. Introducing students to the overarching conservation laws helps develop the analogy that ties the different disciplines together under a common umbrella of system energy. In this book, we use the superhero cast to present the effort-flow analogy and its relationship to the conservation principles of mass, momentum, energy, and electrical charge. We use a superhero movie script common to mechanical, electrical, fluid, and thermal engineering systems to illustrate how to apply the analogy to arrive at governing differential equations describing the systems' behavior in time. Ultimately, we show how only two types of differential equation, and therefore, two types of system response are possible. This novel approach of storytelling and a movie script is used to help make the mathematics of lumped system modeling more approachable for students.

KEYWORDS

mathematical modeling, systems dynamics, transport modeling, lumped system analysis, engineering mechanics, systems modeling, modeling approximation, energy, storage, effort, flow, multi-disciplinary systems

Contents

Preface

If I make a mark in time,
I can't say the mark is mine;
I'm only the underline
Of the word.
Like everybody else, I'm searchin' through
All I've heard.

Cat Stevens
"Tuesday's Dead"

There is a transparency to my accumulated writing. When I look deep
beneath my declarations, I see the underlying thoughts of others. I
realize now how much of what I have said is neither original nor unique.
Thought is forever being revived, recycled and renewed.

Robert Fulghum
Words I Wish I Wrote

The technical content in this book is based on disciplinary physics whose mathematical modeling is well-known. The overarching concepts of effort and flow variables have been presented before in a variety of ways [6–8, 18, 19]. Personally, I wish I'd been taught this way of analogical thinking in my undergraduate studies. Only recently was I taken by the power in the analogy when tasked to teach a course in systems dynamics. In the course of teaching, I developed a story to accompany the analogy. What is offered here is this story. The mathematical relations are not new, but the story is. Like Cat Stevens and Robert Fulghum, I still find value in this interpretation of "words said before." As the physicist Joanne Lavvan admits in her interview for the book *Einstein's God* [17], "I have not changed the facts; I've only changed the approach to the facts."

THE LANGUAGE OF MATHEMATICS

> Schooling, Frey asserts, discriminates against right-brained functions in favor of left-brain functions. Analogical thinking should be done BEST by right-brain-dominant individuals, but transport processes are often taught in an abstract, mathematically oriented manner. Thus, people who should be best able to understand transport process applications must struggle to learn them in the abstract.
>
> Arthur T. Johnson
> *Biological Process Modeling: An Analogical Approach*

Mathematics is the language of modeling. Richard Feynman has called it "the language Mother Nature speaks" [5]. Therefore, it does no good to try to understand her without it. In the business of mathematically modeling material behavior, it turns out that polymer transport of embedded fibers, stresses in dry, densely packed granular materials, and anisotropy in crystalline metals have something in common. Mathematical models for all of these physical phenomena share a common mathematical formulation based on the discipline-specific underlying physics. The ultimate commonality between different physical systems is how they are represented mathematically. What can make studying these fields daunting is the level of abstraction in the mathematics. This mathematics can seem cumbersome, but it is also the single underlying storyline, the common thread for which each of the individual applications is but one manifestation. Mathematics can be like the DNA that is common to two people who are more alike than they appear.

In using mathematics to model, we draw a unique picture of what is inherently similar about distinct scientific disciplines under a wide modeling umbrella. Previous treatments have successfully applied the principles of mathematical modeling to draw the boundaries of this umbrella. But mathematical abstraction has often kept the umbrella at bay for those who think less "left-brained." In today's digital world, more and more is done on our behalf by models and simulations entrusted to the computer and crunching "big data."

> Students don't understand numbers as well as they once did. They rely on the computer's perfection, and they are unable to check its answers in case they type the numbers in wrong. Perhaps our society will decide that the average person does not need to understand numbers and that we can entrust this knowledge to an elite caste (the computer) [but either way] there is a catch. In order to say anything about the universe with mathematics, we have to construct a mathematical model. And models are always imperfect. They always oversimplify reality, and every mathematical model begins with assumptions. Sometimes we forget these are only assumptions. We fall in love with our models. Major trauma ensues when we have to modify or discard them.
>
> Dana MacKenzie
> *The Universe in Zero Words*

Dana MacKenzie may be right that we are possibly moving to a world where mathematics may be the machine behind the curtain. But engineers will still have to build, maintain, and ultimately understand the machine. So, math matters! Ultimately what is essential for today's engineering student is to understand the implications of mathematical simulation performed on their behalf. How that is done is not necessarily the end of the story, but it may be finding the path of least abstract resistance. We ultimately need a way to introduce the mathematics at an appropriate level for new learners. We too often trudge through a nest of complexity trying to find the kernel of wisdom that excites. Complexity is often left to "the experts to explain." The problem is we don't often enough pull it off. Fortunately, complex systems have always been, on some level, simplified through the telling of stories.

THE LANGUAGE OF EXPERTS

> Students are challenged by important aspects of engineering that can seem obvious and easy to experts, the so called "expert-blind-spot" which can impede effective classroom instruction.

> Susan Singer and Karl Smith
> *Understanding and Improving Learning in Undergraduate Science and Engineering*

Singer and Smith [13] make a salient point: that experts have too often forgotten more than students have yet to learn. We're so far into the forest, we may have forgotten how to describe the trees. The reason some experts fail to communicate is that they've been trained to talk in jargon and unnecessary precision which begets complexity without understanding. This sentiment is passionately outlined by Tyler DeWitt, an MIT doctoral student in microbiology and high school teacher: that good science communication can cut through exhaustive detail by telling a good story.

> In the communication of science, there is this obsession with seriousness. Science communication has taken on this idea I call the tyranny of precision where you can't just tell a story. Good storytelling is not about detail; it's all about emotional connection! We have to convince our audience that what we're talking about matters by knowing which details to leave out so that the main point still comes across! The great architect Mies van der Rohe said **"Sometimes you have to learn to lie in order to tell the truth."** I understand the importance of detailed, specific scientific communication between experts. But not when we're trying to teach young learners. (In this case) leave out the seriousness, leave out the jargon, leave out those annoying details, and just get to the point! Make me laugh. Make me care. How should you start? How about saying "Listen let me tell you a story"?

> Tyler DeWitt
> TED Talk: Hey Science Teachers, Make It Fun!

So we set out to tell a story. A story where animation, characters, roles, and a script offer a less formal introduction to the common story of energy storage and loss. The way around abstraction is through metaphor and analogy.

THE IMPORTANCE OF TRIANGULATION

It is of first rate importance that you know how to **"triangulate"** – that is, to know how to figure something out from what you already know.

R.P. Feynman
Tips on Physics

Analogical reasoning is based on the brain's ability to form patterns by association. A new idea is compared to an idea that is already well-understood. The brain may be able to understand (these) new concepts more easily if they are perceived as being part of a pattern.

Jonah Lehrer
How We Decide

Educators can help students change misconceptions by using "bridging analogies" that link students' correct knowledge with the situation about which they harbor false beliefs. Using multiple representations in instruction is one way to move students to expertise.

Susan Singer and Karl Smith
Understanding and Improving Learning in Undergraduate Science and Engineering

The common theme of bridging disciplines shines forth. This can be accomplished powerfully by employing analogies. An emphasis on analogical thinking is adopted throughout this book. The concepts are not new. Only the presentation. There is a common story, "a single script to essentially the same movie." The movie can be set in a variety of stages: electrical current flow, fluid mass transport, heat flow, and momentum transfer. In each of these distinct applications, we are essentially watching remakes of this same underlying movie: same script, same characters, but different actors playing the roles. These different actors bring their own nuanced interpretation to the specific characters they play.

If you've ever seen a re-make of an old movie, you've experienced this sort of thinking. You've seen the story told before through the eyes of one director and a specific cast of actors. In what follows, our pedagogical approach is simply to view the common script through the eyes of four distinct casts. We'll see that the story told is the same, but each cast brings its own distinct feel to the common script. Also, as is the case whenever one is presented with two tellings of essentially the same story, we tend to prefer one cast. People often relate all other interpretations to this favorite telling.

THE CAPTAINS OF ENERGY STORY

> Storytelling provides a method for scholarly discourse in engineering education to make implicit knowledge more explicit, promote reflective practice, and provide entry points into a community of practice.

> C.J. Atman, et al.
> *Enabling Engineering Student Success*

This book uses storytelling to unify the concepts that underlie transport modeling, and make that modeling come alive. In *IMAGINE: How Creativity Works*, Jonah Lehrer [9] describes how such a premise can dramatically awaken the reader:

> Our breakthroughs often arrive when we apply old solutions to new situations. The best way to understand this conceptual blending is to look at the classic children's book *Harold and the Purple Crayon*. The premise of the book is simple: Harold has a magic crayon. When he draws with this purple crayon, the drawing becomes real. If Harold wants to go for a walk, he simply draws a path with his crayon. *But here's the twist that makes Harold and the Purple Crayon (so) engaging*: it blends together two distinct concepts of the world. Although the magic crayon is a fantastical invention Harold still has to obey the rules of reality. When Harold draws a mountain and tries to climb it, gravity still exists in the crayon universe. The book is a delicate balance of the familiar and the fictional.

> Jonah Lehrer
> *IMAGINE: How Creativity Works*

One of the problems with math is that we learn to speak the language on time scales that are not always aligned with our understanding of unifying physical concepts such as energy. Energy is a great unifier of discussions on physical systems, but has not always been exploited as the storyteller that it can be. Energy illustrates a common pattern in each story. In this book, you will be introduced to the Captains of Energy who are at work in engineering systems that are excited by a world outside of themselves, a world controlled by Father Force. Father Force will deliver energy to the system. The Captains will play a game of catch with the imparted energy, a game of monkey-in-the-middle where the Evil Dr. Friction eats away at the energy cache as it is exchanged between Captains Potential and Kinetic Energy again and again. The familiar and fictional are used to unify the mathematical abstraction in an exercise in conceptual blending. The purpose is to convince you that there are only three characters, four casts, one script, and only two equations you need to understand. The purple crayon is tied to reality and made familiar in an attempt to foster longer lasting learning. We script the movie and screenplay with the different actors that appear on the mechanical, electrical, fluid, and thermal stages. One important result of thinking in this way is that you can learn that "breadth at the expense of depth" has inherent advantages for life-long learning. The ability to see how "different things look alike" will equip you with the tools that allow you to adapt to other applications whose underlying physics may be distinctly different, but whose mathematical formulation you have "already seen" before.

OUTLINE OF THE BOOK

Chapter 1 addresses the overarching analogy of all systems variables as belonging to one of two categories:

1. Effort variables and

2. Flow variables

We introduce characters that represent the three key system elements in any transport system: inertia, stiffness, and friction. Thereby, we cast several simple systems in this analogical framework and set the stage for the analogy's universality among separate engineering disciplines. We summarize well-known and essential mathematical relations that correspond to each of the system elements and their respective characters. We introduce the idea that there are separate casts of players in each engineering discipline, but they play the same three roles of the system elements. The script is, in this sense, always the same. Only the actors playing the roles are different. As when any movie is cast with different actors, the same script, when played out, can have a quite different feel, but the storyline remains unchanged.

In Chapter 2, we use the mathematical relations for system elements directly in a conservation principle resulting in a governing differential equation. We provide an example of how this is accomplished for an electrical system, as this is most often the discipline to which all others are made analogous.

In Chapter 3, we illustrate several examples of electrical systems and derive their respective governing differential equations. We examine several possible systems, but stress the procedure more than the system specifics. We do this to emphasize that the specifics can be viewed as incidental. Here, we provide reasoning for students to understand when governing equations will be first order and second order. We also introduce the notion of a normalized form of these equations and their solutions.

Chapter 4 presents the mechanical analog of systems similar to those examined in Chapter 3. Actors in the mechanical cast are presented and their roles in specific systems are offered as examples. We present single and dual energy character scripts that result in first and second order differential equations, respectively.

In Chapter 5, we exploit linearity to find solutions to the normalized governing differential equations in the time domain. We offer an examination of dimensionless solutions as a means to illustrate the concept of a master curve that cements the analogy mathematically. We present the forms of master curves for first and second order systems and set the stage for analogies in fluid and thermal systems.

In Chapter 6, we present classical solutions for systems in steady state that are excited by harmonic loads. Classically referred to as system response in the frequency domain, solutions are obtained via use of Laplace transforms and sinusoidal transform functions. It is typical to see these solutions already in dimensionless form rendering total system solutions that are entirely dimensionless. We explain why casting models in dimensionless form is serendipitous for predictive capability.

In Chapter 7, we present the system analogy for fluid and thermal systems. We illustrate several examples of where first and second order systems arise and the nonexistence of second order thermal systems.

Throughout this book, our intention is to provide an analogous procedure whereby students can see that deriving governing differential equations is a task accomplished always in the same manner, independent of the system's discipline. In the Chapter Activities following Chapters 2–7, we present a small series of applications whereby the analogy can be used to construct equivalent systems that should now "look familiar." We hope this belies a complexity that is born of specific detail, a detail which we argue does not actually exist when one approaches the mathematical model from the perspective of a common movie script merely played out by new and different casts of actors.

Vincent C. Prantil and Timothy Decker
January 2015

Acknowledgments

We greatly appreciate the contributions of Drs. John E. Pakkala and Hope L. Weiss at the Milwaukee School of Engineering for their meticulous reviewing, proofing, and vetting of this manuscript. Their perspectives bring a clarity and consistency to the presentation that it otherwise might not have found. We are grateful to them for providing a review of an earlier draft of this work which helped us to polish and refine many details.

From Vincent C. Prantil: I wish to dedicate this book to my uncle and life-long mentor in all things academic, Dr. Carl Calliari, retired professor of Education at Rowan University. I also wish to thank the enormous vat of patience exhibited by my partner in life and crime, Laurna, and my children, Carmen and Lorin. Their support, laughter, and love continue to carry me through my journey with more encouragement, enthusiasm, and sanity than it otherwise would possess. They have unselfishly encouraged and supported the many adventures in my calling as a teacher. I would like to thank my parents, Dolores and Joseph Prantil, for rearing me in a home with much room for laughter and looking at the world in unconventional ways. They let me find my own way and have always been there to support even the craziest of ideas.

I am grateful for the likes of Steven Strogatz, Michael Guillen, and Bill Nye of Cornell, along with Tyler DeWitt of MIT for their testimony to the art of writing beautifully about science and to the pedagogical power in making science fun. I dedicate this book to *my* students who adopted the energy characters as routes to analogical peg points in the mind's eye. My students doubt, prod, question, and keep me young. We travel through the forest together. I am grateful to Sandy Haeger and her coterie at the One Way Cafe in Wauwatosa, Wisconsin. Sandy weekly allowed me to nurse a bottomless cup of coffee and a hard roll, the smallest tab in the Midwest, while penning these pages and perusing Tim's drawings. I am blessed to have been provided with an amazingly understanding publisher in Joel Claypool and the ever accommodating editor, Andrea Koprowicz, whose encouragement and upbeat demeanor saved many of my faltering moments. Finally, I am forever grateful to my Creator who blesses me every day with a mysterious mix of skepticism, faith, failure, humility, humor, energy, and imagination. *Ego adhuc cognita.*

From Timothy Decker: I dedicate this book to my son Evan, a constant source of support, strength, and hockey. He centers me and makes me remember daily what is important in life. I also dedicate this book to my many students who keep me young, and keep me guessing, laughing, and learning.

Vincent C. Prantil and Timothy Decker
January 2015

CHAPTER 1

If You Push It, It Will Flow

Lenny: "What makes things move, George?"
George: "Forces do, Lenny."
Lenny: "What makes things stop moving, George?"
George: "Forces do, Lenny."

Leonard Susskind and George Hrabovsky
The Theoretical Miminum:
What You Need to Know to Start Doing Physics

At first glance, it is not often evident that individual disciplines in the physical sciences exhibit a fascinating similitude. That is, behaviors in distinct fields share a unifying theme. For instance, a voltage drop across a circuit causes charge or current to flow. Similarly, a temperature difference causes heat to flow from hot to cold. The windfall for engineers is that the mathematical models for these transport processes, either for current or heat, are identical! Richard Feynman has said that "mathematics are the eyes with which we see physics" [3, 5]. To the more physically inclined, this may appear to be placing the cart ahead of the horse. But when we view the world this way, models allow us to "see" a unifying theme that underlies what we physically observe. Mother Nature, in her sense of orderliness, has chosen to sing a similar song in different keys. The music is mathematics [2]. But mathematics can be a double-edged sword. While it can help us to see patterns and maybe even search for physical insight through patterns, it can be abstract and elusive for the new learner with less experience using their newly acquired tools of calculus.

Here, we define a movie script that has only four character roles. These will be a character putting energy into the system, two characters who store energy, and an energy eater. These roles will be played by a new and different cast in each discipline (the electrical, mechanical, fluid, and thermal worlds). When a movie is remade with new actors portraying the characters, often people will take a liking to one cast over another. In other words, one particular cast of actors bring the screenplay to life in a particularly more meaningful way for them. So the relationship between voltage and current above is analogous to an identical relationship between temperature and heat flow. Often engineers with a propensity for viewing the world "electrically" can translate a thermal system into an equivalent electrical one for the purposes of understanding "the movie" with a new and different cast. The reason this is so is because there is a common framework in which current and heat flow may be cast where the characters are the same; they are merely portrayed by different actors. This analogical thinking is a formidably powerful tool for fostering learning.

1.1   THE EFFORT-FLOW ANALOGY

All learning is by analogy.

Albert Einstein

No set of engineering principles is more useful or pervasive than the concepts of effort variables and flow variables. By analogy, these can be applied to almost any situation involving transfer of something from one location or situation to another.

A.T. Johnson
University of Maryland, College Park

The substrate of analogical thinking involves recognizing a commonality between what, on the surface, may initially appear to be unrelated. For instance, the flow of mass, momentum, heat, and electrical charge are not as independent as they may appear at first glance. In fact, a powerful unifying theme or analogy exists linking the transport models in these otherwise distinct disciplines.

Figure 1.1: A force applied to a mechanical system causes motion to occur. Force must continually be applied in the presence of friction if motion is to continue.

Effort variables represent the force-like quantities, forces in and on a system. Flow variables are quantities that change in response to the applied effort. The effort and flow are called conjugate pairs because they are necessarily married in a description of work and energy. Consider the example of a force applied to a block along a frictional surface. If there is sufficient force, the block will move. The block is a system characterized by its inertia and the friction between the block and the floor. A character we will call Father Force provides an externally applied effort to the block. Father Force lives in a world outside of the system. The external force or effort he supplies, if high enough to overcome the friction force, will cause a change in the block's velocity or flow.

The force on the block and the resulting motion cannot be specified independently, i.e., there is an explicit relation between these two quantities. We can associate motions with requisite forces or, just the same, forces with the ensuing motion. While causality is, in some sense, in the mind of the observer, we can agree from this point on that a force applied to a system causes motion to take place. It is these quantities of force and subsequent motion that will form the basis for an elementary analogy. Consider now an electrical analog to this mechanical system. If you place a voltage difference across a resistor, current will flow through the resistor. For a known amount of resistance, you cannot specify the voltage difference applied and the resulting current independently. They are related. The electrical voltage difference acts like a net force. This net force pushes electrical charge through the resistor. The resistor represents an electrical analog to friction, if you will. And the current is a rate of change of electrical charge with time, just as the velocity of the block is a rate of change of displacement with time. What remains the same is that when you place an effort difference or a net effort **across** a system, flow occurs **through** the system.

> *Of Special Note*
>
> *In any transport process, a difference in an effort variable across a small region of a system drives a transport or flow of some quantity through the small region.*

So, a force difference or net force across a mass will cause a change in its momentum. A difference in electrical potential (or voltage) causes current to flow. A temperature difference causes heat to flow while a pressure difference causes fluid to flow. In a unifying template, force, voltage, temperature, and pressure play analogous roles. They are the effort driving the flow of, respectively, momentum, (electrical) charge, (heat) energy, and mass. These are the four quantities that are classically conserved or balanced in all systems. These are four quantities you can neither create nor destroy. Effort always drives flow. And what flows is usually related to whatever is conserved. Learn to think this way and almost everything you will learn in engineering will abide by this same set of rules wherever transport or dynamics are involved.

We list the conjugate effort and flow variables for the four separate disciplines in Table 1.1. Here, the four disciplines have fostered models that describe how mechanical momentum, fluid mass, electrical charge, and thermal heat flow under the influence of force, pressure, voltage, and temperature differences respectively.

In the course of your education, you may come across the nomenclature of a generalized force. A simple description in the current context is that a generalized force acts through a gen-

Table 1.1: Effort and flow variables used to describe transport of momentum, mass, heat, and charge

Discipline	Effort	Flow
Electrical	Voltage	Current
Mechanical	Force	Velocity
Fluid	Pressure	Mass Flow Rate
Thermal	Temperature	Heat Flow Rate

eralized displacement to produce work [16]. In a mechanical system, forces act through displacements to do work. In rotational mechanical systems, torque acts through an angular displacement to perform work (see Table 1.2). Note the units of force multiplied by displacement, e.g., N-m or joules, J, is the same as the product of torque and angular displacement, Nm-rad or N-m or J, units of work (and energy). In electrical systems, the product of voltage and charge is given by the product of volts and coulombs. By definition, this product is also measured in joules, J. We have chosen to associate flow with the time rate of change of a displacement-like quantity, e.g., velocity, angular velocity, or current. As such, we will work with the following convention: effort is a generalized force, while flow is the derivative of a generalized displacement. The product of effort and flow will result in power or the rate at which work is performed on or energy is input to a system.

Table 1.2: Concept of generalized force and motion in mechanical systems

Discipline	Effort	Flow
Mechanical	Generalized Force	Generalized Motion
Translational	Force	Velocity
Rotational	Torque	Angular Velocity

Of Special Note

Because we follow the flow of a conserved quantity, most often the flow variable is the time rate of change of the conserved quantity.

1.1.1 SYSTEM ELEMENTS

The screenplay of the transport process movie is written in terms of energy which is always conserved. As we will see soon, the concept of conservation plays a critical role in modeling. Since all real systems involve losses in energy, it would be more correct to say that energy is always balanced. The balance is composed of two types of stored energy pitted against the eventual losses. So our movie has two types of characters or elements: those that store energy and those that dissipate energy. Further, there are two elementary storage characters: those that store potential energy and those that store kinetic energy.

Storage Elements

The key role of the system elements or components in modeling is that they represent explicit relations between the effort and flow. A system element will be portrayed by a character in our movie. Any transport process can, at any moment in time, store energy by virtue of its effort variable or its flow variable. We call energy stored by virtue of a system's effort variable potential energy. Any system element that stores potential energy will play the role of Captain Potential Energy. This energy is locked inside a system by way of an effort difference that can be relaxed to allow the energy to be released in a form evidenced by the system's flow variable. Energy stored by virtue of a system's flow variable is kinetic energy. Any system element that stores kinetic energy will play the role of Captain Kinetic Energy. In what follows, we will write mathematical expressions for the energy storage that will have analogs in each discipline of study. They will always look the same. To plant the analogy, we choose the electrical and mechanical disciplines to demonstrate examples of the system elements or characters. We will also use these disciplinary examples to attempt to shed light on "how" potential and kinetic energy are stored and "who" stores them.

Potential Energy Storage Elements   Those elements of transport that store potential energy do so by virtue of building up an effort difference that can be released to perform useful work. In these cases, flow is always proportional to a time derivative of effort:

$$FLOW \propto \frac{d}{dt}[EFFORT] \tag{1.1}$$

where the proportionality constant determines the specific amount of flow released upon relaxation of an effort difference or the capacity of the process to perform work. As such, we term this constant the system capacity or capacitance, C.

Of Special Note

Energy stored by virtue of stored differences in effort is potential energy. Characters that store potential energy follow the equation:

$$FLOW = C \frac{d}{dt}[EFFORT] \tag{1.2}$$

that defines the character's capacitance.

Recall that in electrical circuits, capacitors are system elements that store energy through voltage differences across dielectric plates. Upon discharge, a flow of charge or current is released. For this process:

$$i(t) = C\frac{dV(t)}{dt} \tag{1.3}$$

Analogously, we may ask which system element stores potential energy in a mechanical system. We typically recall from elementary mechanics that this is a spring. Potential energy is stored by virtue of a stored effort or mechanical force in the deformed spring. Recall that the force and displacement are related by Hooke's law for simple, linear springs.

$$F = kx \tag{1.4}$$

No differential relation is evident, so let's examine our storage a bit more closely. Recall that the flow variable is velocity, the time derivative of displacement. In fact, flow variables are often related to conserved or balanced quantities. In mechanical systems, momentum is balanced. In systems where mass is constant, this implies that velocity is the appropriate flow variable when linear momentum is conserved. Using the definition of velocity as the time derivative of displacement relates velocity to a time derivative of force:

$$x(t) = \frac{1}{k}F(t)$$
$$\Rightarrow v = \frac{dx(t)}{dt} = \frac{1}{k}\frac{dF(t)}{dt} \tag{1.5}$$

Then, by analogy, the mechanical capacitance is given by the reciprocal of the spring stiffness:

$$C_{MECH} = \frac{1}{k} \tag{1.6}$$

Generalizing, by analogy, a transport process exhibits a capacitance given by:

$$C = \frac{1}{EFFORT} \int (FLOW)\, dt \tag{1.7}$$

When the energy is stored by virtue of effort, it is potential energy. Energy is given by an integral of power expended in a process. Thereby, the potential energy stored in a capacitor would be given as:

$$\int V(t)i(t)\, dt = \int \left(\frac{1}{C}\int i(t)dt\right) i(t)\, dt$$
$$= \int \left(\frac{1}{C}q(t)\right) i(t)\, dt = \int \left(\frac{1}{C}q\right) dq = \frac{1}{2}\left(\frac{1}{C}\right) q^2 \tag{1.8}$$

where the definition of current is:

$$i(t) = \frac{dq(t)}{dt} \tag{1.9}$$

Once again, analogously in a mechanical system, the potential energy stored by a spring by virtue of the force within it is given as:

$$\int F\upsilon \, dt = \int (kx)\upsilon \, dt = \int (kx) \, dx = \frac{1}{2}kx^2 \tag{1.10}$$

You may recall from your elementary physics courses that this is the expression for potential energy in a deformed spring. To begin our energy story, any system element who stores potential energy is Captain Potential Energy. He possesses energy by virtue of effort or the force contained in his springs!

Figure 1.2: Captain Potential Energy stores energy by virtue of effort in his compressed springs. The containment vessel for the effort is the stiffness or capacitance. Captain Potential Energy is distinguished by his possession of a system's capacitance.

Kinetic Energy Storage Elements   Those elements of transport that store kinetic energy do so by virtue of their flow variable. As a result, effort differential is related to a time derivative of flow:

$$EFFORT \propto \frac{d}{dt}[FLOW] \tag{1.11}$$

Where the proportionality constant determines the specific amount of effort difference required to cause the prescribed rate of change of flow. This term is referred to as the system inductance, L.

Of Special Note

Energy stored by virtue of stored flow is kinetic energy. Characters that store kinetic energy follow the equation:

$$EFFORT = L \frac{d}{dt}[FLOW] \tag{1.12}$$

that defines the character's inductance.

Recall that in electrical circuits, inductors are system elements across whose terminals a voltage drop is related to a time rate of change of current.

$$V = L\frac{di(t)}{dt} \tag{1.13}$$

Analogously, we may ask which system element stores kinetic energy in a mechanical system. We typically recall from elementary mechanics that this is the mass or inertia of the system. This comes naturally from Newton's Second Law relating the net force on a mass to its time rate of change of linear momentum. When mass is constant, the rate of change of momentum is proportional to the mass's acceleration.

$$F = ma = m\frac{dv(t)}{dt} \tag{1.14}$$

Thus, when a mass exhibits some non-zero speed, it possesses kinetic energy by virtue of its speed and in proportion to its mass or inertia. If there were no mass, there would be no entity to have speed! In this interesting way, we can learn to say that the mass stores the kinetic energy in the form of its speed. Because the mass stores the kinetic energy in a mechanical system, we can say that

$$L_{MECH} = m \tag{1.15}$$

Generalizing, by analogy, a transport process exhibits an inductance given by:

$$L = \frac{1}{FLOW}\int (EFFORT)\, dt \tag{1.16}$$

With energy being an integral of power, we can work in terms of the flow variable to define the kinetic energy:

$$\int V(t)i(t)\, dt = \int \left(L\frac{di(t)}{dt}\right)i(t)\, dt = \int (Li)\, di = \frac{1}{2}Li^2 \tag{1.17}$$

Once again, analogously in a mechanical system

$$\int F\upsilon(t)\,dt = \int \left(m\frac{d\upsilon(t)}{dt}\right)\upsilon\,dt = \int (m\upsilon)\,d\upsilon = \frac{1}{2}m\upsilon^2 \tag{1.18}$$

This may look familiar to you as the kinetic energy in a moving mass. Continuing our energy story, any system element who stores kinetic energy is Captain Kinetic Energy. He possesses energy by virtue of flow or the velocity associated with his mass or inertia!

Figure 1.3: Captain Kinetic Energy stores energy by virtue of his speed. The containment vessel for the speed is the inertia or mechanical inductance. Captain Kinetic Energy is distinguished by his possession of a system's inductance.

So let's remember that mathematically, both storage elements (or characters) relate flow or effort to the derivative of the other. Differential relations imply energy storage.

Of Special Note

Differential mathematical relationships for system elements imply energy storage.

Dissipative Elements

In dynamic transport, dissipative elements relate flow and effort strictly algebraically. Algebraic relations imply energy dissipation. Resistive elements, in principle, play a part in every disciplinary story. Flow experiences resistance under the action of any difference in effort that drives it:

$$EFFORT \propto FLOW \tag{1.19}$$

Here the proportionality constant determines the specific amount of resistance that must be overcome by a given net effort to drive a given amount of flow. This term is referred to as the system resistance, R.

> *Of Special Note*
>
> Characters that dissipate energy follow the equation:
>
> $$EFFORT = R * FLOW \tag{1.20}$$
>
> that defines the character's resistance.

where $*$ indicates multiplication.

Figure 1.4: The Evil Dr. Friction dissipates energy as flow occurs under the driver of an effort difference across some resistive element. The Evil Dr. Friction is distinguished by his possession of a system's resistance to flow.

Recall that in electrical circuits, resistors are system elements across whose terminals a voltage drop is proportional to the current flowing through it as prescribed by Ohm's law:

$$V = iR \tag{1.21}$$

Similarly in a mechanical system, force can be applied to produce motion by overcoming the effects of friction. For example, when viscous forces oppose the motion, a good representation of the force required to overcome this resistance is given by:

$$F = bv \tag{1.22}$$

where clearly, by analogy,

$$R_{MECH} = b \tag{1.23}$$

Of Special Note

Algebraic mathematical relations imply energy dissipation. Often, these dissipative relations bear someone's name, e.g., Ohm, Fourier, Newton, Toricelli, etc.

Generalizing, by analogy, a transport process exhibits a resistance given by:

$$R = \frac{EFFORT}{FLOW} \tag{1.24}$$

The energy "eaten by" any resistive element is equivalent to the work done by the dissipating agent, e.g., friction in a mechanical system. When energy is "eaten" it is no longer available to be stored in potential and/or kinetic forms. We say it is effectively "lost." The lost or dissipated energy is quantified by the work done by the force or effort across the element:

$$\int \frac{dW}{dt} \, dt = \int dW = \int \left(i(t)^2 R \right) dt = \int (V(t)i(t)) \, dt = \int V(q) \, dq \tag{1.25}$$

Analogously in a mechanical system, the lost energy is given by:

$$\int \frac{dW}{dt} \, dt = \int dW = \int F_{FRICTION} v \, dt = \int F_{FRICTION} dx \tag{1.26}$$

You may recall from undergraduate engineering dynamics that this expression is equivalent to the work done or energy dissipated by friction acting on a moving mass. As energy is transported and exchanged between potential and kinetic forms, resistive agents essentially steal part of the transfer. There is a balance between energy transferred and energy lost. The resistive element acts to transform energy to a form not useful by the particular system, i.e., resistors transform useful electrical energy to heat, a loss by-product of current flow in a real circuit. Similarly, friction in mechanical systems steals energy, transforming it to sound and heat, no longer useful for producing motion. The Evil Dr. Friction is the character who irreversibly robs energy in a system as it is being released from either potential or kinetic forms and in any transfer between the two.

1.1.2   THE ENERGY BALANCE PRINCIPLE

In any given system, transport occurs when effort drives flow of some quantity. By defining a control volume into and out of which flow occurs, one can create a balance of any quantity, Q, by stating that

$$\dot{Q}_{IN} - \dot{Q}_{OUT} = \dot{Q}_{STORED} \tag{1.27}$$

where $\dot{Q}_{IN} = \frac{dQ_{IN}}{dt}$. As we will see, this simple balance principle is the precursor to every differential equation governing the transport of conserved or balanced quantities. Generally, input and output transport will require overcoming some resistance to flow with the net inflow resulting in storage.

CHAPTER 2

Governing Dynamics

Governing dynamics, gentlemen; it's all governing dynamics

John Nash

In all of science we take certain axioms as given and proceed to model behavior from there. One such premise is that there is a collection of quantities whose behavior is governed by principles of conservation and balance. Among these are mass, momentum, energy, and electrical charge. We accept that these quantities can neither be created nor destroyed. Therefore, the amount of any one of these quantities is a function of how much you begin with and how much is either transported to you or from you. There can be external sources and sinks and repositories where the quantities can be stored. To write a statement that balances any conserved quantity at a point, we isolate an infinitesimally small volume with inlet and exit windows through which our quantity of choice can be transported in and out.

It is perhaps best at this point to proceed by example. In an attempt to be consistent with other treatments of the effort-flow analogy, let's consider a volume, e.g., a bank vault, into which money may enter and exit through different ports in the volume boundary, e.g., the bank doors. In order to introduce the parameter of time, imagine that dollars enter the bank at a rate of $\dot{Q}_{IN}$ dollars per day. Say a different amount may be flowing out of the bank at a rate of $\dot{Q}_{OUT}$ dollars per day. A balance principle is as simple as tallying how much money enters vs. how much exits. If the amount entering is greater than the amount exiting in any given interval of time, there is a net accrual and the amount of money in the bank increases over time. In other words, there is a net amount of storage of money in the bank. Contrarily, if the amount exiting exceeds the amount entering in any time interval, the amount of money in the bank will decrease. One can then conclude that the amount stored in this time interval is a negative value, i.e., there is a net loss of money when $\dot{Q}_{STORED} < 0$ over this time interval. So the statement of balance (in rate form) is simply

$$\dot{Q}_{IN} - \dot{Q}_{OUT} = \dot{Q}_{STORED} \tag{2.1}$$

In the absence of a storage mechanism, the amount of the quantity already stored in this volume remains constant and we say this quantity is conserved. When this is the case, any net inflow must be accompanied by an equal amount of outflow.

Figure 2.1: A repository for a balanced quantity that allows inflow and exit from it through virtual windows and storage on the interior.

2.1   DERIVING A GOVERNING DIFFERENTIAL EQUATION

We will need to explore some of the more germane properties of differential equations here to establish the utility of our analogy. When we let the size of the volume into which a balanced or conserved quantity flows shrink to a point, the balance or net storage principle becomes a differential equation governing the amount of the quantity present at that point at any given moment in time. Once again, to make the mathematics more approachable, let's proceed by example. Let's start with a simple story of a passive electrical circuit that contains an energy dissipator, the resistor, connected in series with an electrical potential energy storage device known as the capacitor as shown in Figure 2.2. At some point in time, a DC battery is connected across the circuit. In our cartoon version in Figure 2.3, Father Force represents an excitation from the outside world, an externally applied effort. We can often understand system behavior by associating this force with a driving external agent from the outside world perturbing the system. Here, the agent of the outside world, the battery, imposes a voltage difference across the circuit. In our story, we call this agent of the outside world Father Force because he represents an externally applied effort difference. He hurls electric charge at the circuit, our system. This electrical voltage difference drives electrical charge at some rate (known as current) through the resistor. The resistor is an energy dissipator, the Evil Dr. Friction in our cartoon. His snake "eats charge" and thus electrical energy. He steals it from the system. This electrical energy will be lost mostly in the form of heat. What is left exits the resistor and can be stored across the plates of an electrical capacitor. Here we see Captain Potential Energy as the storage agent in the capacitor.

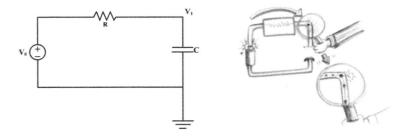

Figure 2.2: A series RC circuit as typically represented with a standard circuit diagram. A representative volume element of the circuit is examined under the magnifying glass. Given that the circuit is grounded at the lower battery post, a voltage at the prescribed node represents the voltage drop across the capacitor, V_1.

At this point, we should point out that the system is defined exclusively by the system element characters, i.e, one character that eats energy and one that stores energy. The battery is "the outside world." This agent imposes a voltage difference on the circuit that causes current to

Figure 2.3: A electrical effort (voltage difference provided by a DC battery) throws charge flow through a resistor, an energy dissipator, which eats part of the input charge only to have the amount that gets through be stored by a storage element or character. As the voltage difference across the capacitor grows storing electrical potential energy, the voltage drop across the resistor decreases and, along with it, the current in the circuit.

flow. Father Force lives in the outside world and delivers an input to the system whose characters are Captain Potential Energy and the Evil Dr. Friction!

To derive any governing differential equation, we isolate a small part of the system. Consider a point in the circuit between the resistor and the capacitor (the node under the magnifying glass in Figure 2.2). This choice of representative volume element (RVE) is somewhat arbitrary. Since the only charge storage element in our example is the capacitor (and this character is outside of the RVE), the amount of charge per unit time (or current) flowing into the node must exactly equal or balance the current flowing out because there is no means by which to allow charge to accumulate on a wire alone.

This concept must now be rendered mathematically. The current IN must equal current OUT or the current that passed through the energy dissipator must equal that flowing into the energy storage element. Material laws usually govern the inflow and outflow of the balanced quantity. These material laws are hypotheses based on observation and measurement [1, 2, 12, 14].

$$\dot{Q}_{IN} - \dot{Q}_{OUT} = \dot{Q}_{STORED}$$
$$\Rightarrow i_R - i_C = 0$$
$$\Rightarrow i_R = i_C \qquad (2.2)$$
$$\frac{V_O - V_1}{R} = C \frac{d(V_1 - V_{REF})}{dt}$$
$$RC\dot{V}_1 + V_1 = V_O(t)$$

where V_O is the battery voltage and V_1 is the voltage drop across the capacitor.

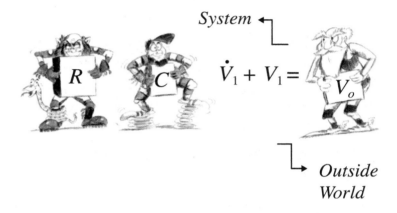

Figure 2.4: The abstraction of the mathematical governing equation exhibits "sides" belonging to the system and the outside world which acts to excite the system into some manner of dynamic response.

Because the reference voltage is chosen to be grounded, that is zero voltage, we say that this differential equation governs the voltage drop, V_1, across the potential energy storage device or capacitor. It is important to note that the resistor, capacitor, and the interior system voltage, V_1, lie mathematically on one side of the equation while the driver from the outside world lies on the other side of the equation. The left side contains all system parameters and quantities while the right hand side represents a "forcing function" that drives the flow.

$$RC\dot{V}_1 + V_1 = V_O(t) \qquad (2.3)$$

This will be a constant theme in our development. The movie characters and their behavior live "on the left" while the circumstances presented to them by the outside world (and to which they must respond) will lie "on the right" (see Figure 2.4).

2.2   THE FOUR CASTS

Movies are always open to being remade ... I think of it like the James Bond movies. Different actors can play the same role.

Steve Martin

I think the movie business is all movies that you've seen before. Everything's a remake; people want things that are familiar.

Graydon Carter

So the story of the evolution of quantity, Q, over time is governed by the balance

$$\dot{Q}_{IN} - \dot{Q}_{OUT} = \dot{Q}_{STORED} = \dot{Q}_{STORED}^{POTENTIAL} + \dot{Q}_{STORED}^{KINETIC} \qquad (2.4)$$

Here, we must make a statement, though, that while it is allowable, there need not be two storage characters. Recall, the example given in Section 2.1 had only a potential energy storage element in its cast. So the transport processes of interest are those that contain:

1. Dissipative elements and a potential energy storage element

2. Dissipative elements and a kinetic energy storage element

3. Dissipative elements and both potential and kinetic energy storage elements

4. Both potential and kinetic energy storage elements and no dissipative elements

It will turn out that systems with only one type of storage element character in their script are always governed by first order ordinary differential equations in time. Alternatively, systems whose script contains two types of storage element will always be characterized by second order ordinary differential equations in time. These are important characteristics to be aware of before we discuss the nature of their solutions.

The focus of the analogical approach is its power in describing the similitude between systems transporting conserved quantities in four otherwise distinct disciplines of engineering. Dynamic differential equations are statements of how conserved quantities change in time. In electrical systems, we will always balance electrical charge; in mechanical systems, momentum; in fluid systems, mass; and in thermal systems, heat energy. These are summarized in Table 2.1.

2.3   SYSTEM ORDER

In our analogy to a screenplay, we have limited our discussion to two scripts: those associated with first order governing equations and those associated with second order governing equations. While it is seldom said in this way, the order of the governing differential equation is defined as

Table 2.1: Conserved quantities

Discipline	Conserved Quantity
Electrical	Charge
Translational Mechanical	Linear Momentum
Rotational Mechanical	Angular Momentum
Fluid	Mass
Thermal	Internal Energy

the difference between the highest order derivative appearing in the equation and the lowest order derivative appearing in the equation. It will be shown that the system order is the most important determinant of the system behavior. We will have more to say about this as we set about solving these equations in Chapters 5 and 6.

2.4   LINEARITY

It will suffice to say that a differential equation is linear when the system variable and all its derivatives on the left side of the equation, i.e., those associated with the capacitor voltage, V_1, in the example of the previous section, appear only to the first power and there are no transcendental or trigonometric terms on the left side, e.g., exponential functions, natural logarithms, or periodic functions of the dependent variable. While analytical, functional solutions do exist for so-called nonlinear systems, they are mostly rare or difficult. Therefore, solutions to nonlinear systems often require numerical solution techniques. Solutions to governing differential equations are the mathematical representations of physical system behavior. We will concern ourselves with linear systems only in this book. We can use the linear story to help us visualize and understand nonlinear behavior once we master a linear understanding. When appropriate, we can then linearize nonlinear systems to find a simpler story over a limited range of behavior.

CHAPTER 3

The Electrical Cast

An electron's journey through a circuit can be described as a zigzag path that results from countless collisions with the atoms of the conducting wire. Each collision results in the alteration of the path, thus leading to a zigzag type motion. While the electric potential difference across the two ends of a circuit encourages the flow of charge, it is the collisions of charge carriers with atoms of the wire that discourages the flow of charge.

The Physics Classroom

In Chapter 2, our example system was a passive RC circuit, a system whose script contains only two character "types": a potential energy storage character and a dissipative character. In this system, the battery is an agent of the outside world that continually hurls charge through the resistive element that "eats charge" and turns its electrical energy to heat or thermal energy. It is important to note that the energy is not destroyed, but merely transformed to another form that is no longer available as electrical potential causing current flow through the circuit. This represents a loss of so-called electrical energy to other non-useful forms (in terms of hurling electrons through the electrical circuit). The heat in a light bulb is a necessary loss incurred as current flows through a resistive filament which produces heat AND light. Perhaps ironically, the light is a useful "by-product" of the circuit, but, from a purely electrical perspective, it represents a loss of electrical energy that forces the circuit to require constant energy input.

3.1 EFFORT AND FLOW VARIABLES

If you push charge, it will flow. The flow of charge is, by definition, an electrical current. How you push charge is by creating a difference in electrical potential (or voltage). The electrical potential or voltage drop along a portion of a circuit drives the charge to flow from higher to lower electrical potential. Electrical charge can be difficult for mechanical engineers to grasp if we look at the world as driven by external forces that require contact to initiate motion. Electrical charge responds to force-at-a-distance, force fields that are many orders of magnitude larger than, say, gravitational forces in mechanical systems. The reason forces are not always seen as this large is that many, many positive and negative electrical charges end up cancelling each other out. It is only the sparse imbalances in charge that occasionally occur that tip the balance and end up creating a difference in electrical potential. This difference is not an equilibrium state and charges tend to move to reduce this difference. So charge moves in response to the electromagnetic field, a force felt by charge when all charges are not paired up [4]. But it is also a known condition that, like mass and energy, no one can create or destroy charge. Positive and negative charges exist. On the whole, they cannot be created or destroyed, but they can be collected in such states that differences in net amounts drive flow of unlike charges toward one another. All governing equations are based on writing mathematical statements of this conservation of electrical charge. Given that charge is conserved, governing equations of motion arise out of balancing electrical "forces" that drive charge to move toward an equilibrium state.

Table 3.1: Effort, flow, and conserved quantities for electrical systems

Conserved Quantity		Units	Symbol
Charge		Coulombs	q
Variable		Units	
Effort	Electrical Potential Voltage	Volts	V
Flow	Current	Amperes	i

3.2 STORAGE ELEMENTS

All such powered passive electrical circuits can, at most, contain three system element characters. Recall that two of the characters are capable of storing energy, one in the form of potential energy, the other in terms of kinetic energy.

3.2.1 POTENTIAL ENERGY STORAGE CHARACTER

Potential energy storage devices store energy in the form of the effort variable. The electrical cast member who plays the role of Captain Potential Energy is a device that stores a differential of electrical effort or voltage. This is the capacitor.

The potential energy storage character is described mathematically in the same way for *every* disciplinary system. The governing mathematical expression of the storage by virtue of effort is

$$FLOW = C\frac{d(EFFORT)}{dt},$$
$$i_C = C\frac{d(V_1(t) - V_{REF})}{dt},$$

where capacitance is measured in *farads* or ampere-seconds/volt

$$f \doteq \frac{A-s}{V}.$$

Figure 3.1: The electrical potential energy storage character is played by the capacitor.

3.2.2   KINETIC ENERGY STORAGE CHARACTER

Kinetic energy storage devices store energy in the form of the flow variable. The electrical cast member who plays the role of Captain Kinetic Energy is that device that stores energy by virtue of electrical flow or current. This is the inductor.

The kinetic energy storage character is described mathematically in the same way for every disciplinary system. The governing mathematical expression of the storage by virtue of flow is

$$EFFORT = L\frac{d(FLOW)}{dt},$$
$$V_1(t) - V_2(t) = L\frac{d(i_L(t))}{dt},$$

where inductance is measured in *henries* or volt-seconds/ampere

$$H \doteq \frac{V - s}{A}.$$

3.3   DISSIPATIVE ELEMENTS

Energy losses occur at the hand of a dissipative element or a character that "eats energy." The role of the Evil Dr. Friction in the electrical script is played by the resistor.

Recall, the governing mathematical expression of the dissipation is always *algebraic* rather than differential:

$$EFFORT = R * FLOW,$$
$$V_1 - V_2 = Ri_R,$$

Figure 3.2: The electrical kinetic energy storage character is played by the inductor.

Figure 3.3: The electrical energy dissipative character is played by the resistor.

where resistance is measured in *ohms* or volts/ampere

$$\Omega \doteq \frac{V}{A}.$$

Although there is no hard and fast rule about this, the mathematical expression governing energy loss is very often characterized as "someone's law." Here, Ohm's law governs the electrical energy dissipated in a resistor. Because dissipative elements result in energy losses, a requisite effort differential, here the voltage drop, $V_1 - V_2$, is necessary to drive a current, i_R, through the element.

A summary of the electrical cast and the roles they play is given in Figure 3.4 and a summary of the relevant system element relations is summarized in Table 3.2.

Table 3.2: Relevant system element relations for electrical systems

Field	Effort Variable	Flow Variable
Electrical	Voltage	Current
Relation	Form	Analogy
Dissipative Material Property Law	Effort = Resistance x Flow $(V_1 - V_2) = R\,i$	Resistance = Resistance
Energy Storage in Effort Variable	Flow = Capacitance x d(Effort)/dt $i = C\,\dfrac{d(V_1 - V_2)}{dt}$	Capacitance = Capacitance
Energy Storage in Flow Variable	Effort = Inductance x d(Flow)/dt $(V_1 - V_2) = L\dfrac{di}{dt}$	Inductance = Inductance

3.4   SINGLE STORAGE ELEMENT SCRIPTS

Recall, single energy storage scripts are capable of storing only one type of system energy, potential or kinetic. When coupled with an energy dissipating agent, first order ordinary differential equations are the result. These first order equations in time govern the system effort and flow behavior(s). For the electrical cast, the simplest examples are the series RC and LR circuits.

3.4.1   RC CIRCUITS

In the case of the series RC circuit (Section 2.1), electrical energy is provided from "the outside world" by the electrical potential boost of the DC voltage source or battery. Some energy is lost through the resistor, yet enough gets through so that a potential difference or voltage builds up

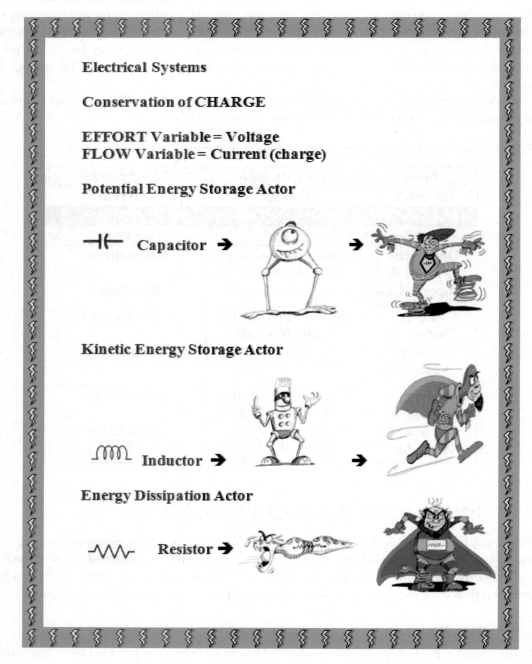

Electrical Systems

Conservation of CHARGE

EFFORT Variable = Voltage
FLOW Variable = Current (charge)

Potential Energy Storage Actor

Capacitor →

Kinetic Energy Storage Actor

Inductor →

Energy Dissipation Actor

Resistor →

Figure 3.4: The electrical cast of characters.

across the plates of the potential energy storage element thereby charging the capacitor through a differential of voltage or effort. Recall, what results mathematically is a differential equation for the voltage stored across the capacitor plates that is linear and first order. Therefore, the resulting differential equation is written for the effort variable.

$$RCV_1 + V_1 = V_O(t)$$

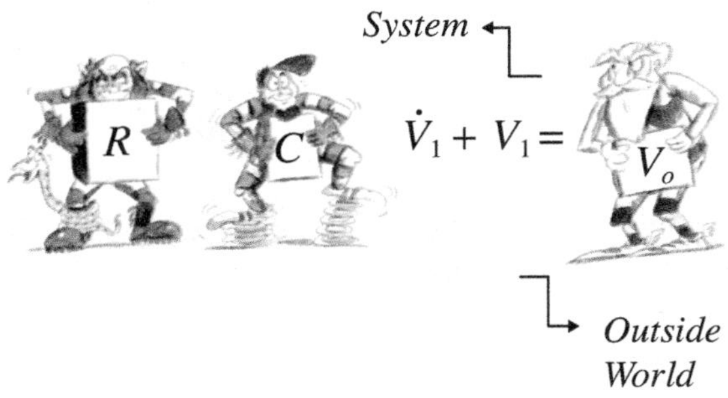

Figure 3.5: The electrical cast of characters playing out a series RC circuit.

Here, the quantity RC possesses units of time.

$$RC \doteq \Omega f,$$
$$RC \doteq \frac{V}{A}\frac{A-s}{V} \doteq s.$$

It is known as the system time constant, $RC = \tau$. Because time is the independent variable and all responses are time histories, it is of primary importance that the system is characterized by this special amount of time. Effort (voltage) and flow (current) will change over time. Because the time constant, τ, enters the governing differential equation explicitly, it flavors the entire system response. How fast or slow effort and flow evolve in time in the system will always be in quanta of time constants. That is, system variables will change explicitly in "chunks" of time of τ seconds. We will learn to talk in these terms. The amount of time it will take for any change to occur in a first order system will be N time constants. The number, N, of course, will depend on what phenomenon we are discussing. We will call the time constant, τ, the system parameter. It parameterizes how fast the system responds to external stimuli.

Electrical systems, it turns out, are "mathematically versatile" in that the resulting ordinary differential equations will as often govern the behavior of the effort variable, V_1, as the flow variable, i. The equation governing the capacitor voltage can be recast as a 1st order ordinary differential equation governing the system current, i.

$$\frac{V_O(t) - V_1(t)}{R} = i(t).$$

Inverting the potential energy storage relation for the capacitor will give an expression for the voltage, $V_1(t)$, which may be substituted into this relation:

$$\frac{V_O(t) - \int \frac{i(t)}{C}\,dt}{R} = i(t)$$

$$i(t)R = V_O(t) - \int \frac{i(t)}{C}\,dt$$

$$RC\frac{di(t)}{dt} + i(t) = C\dot{V}_O(t),$$

where $C\dot{V}_O(t)$ is an equivalent input signal current seen as a forcing function by the differential equation governing the system current. It is associated with the time rate of change of the imposed battery voltage, $V_O(t)$. It is important to note that *the same system time constant, $RC = \tau$*, appears in and characterizes the solution(s) of both differential equations: those governing the capacitor voltage (effort) and the circuit current (flow)!

3.4.2   RL CIRCUITS

We might as easily let the current that passes through the resistor be stored in the form of electrical kinetic energy. The cast member that plays the character storing kinetic energy is the inductor.

Figure 3.6: An electrical effort (voltage difference) drives charge flow through an energy dissipator, the resistor, only to have the amount that gets through be stored by a storage element or character, the inductor, in the form of the flow variable.

We are used to representing this system by a circuit diagram as in Figure 3.7.

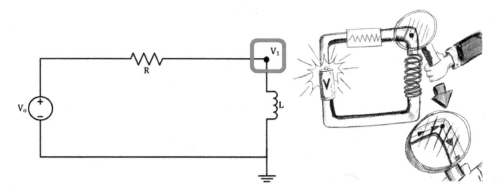

Figure 3.7: An electrical RL circuit as typically represented. Upon applying the battery (an external voltage difference) across the circuit, charge will respond to the electromagnetic force and flow through the circuit.

At this point, we should point out that the *system* is defined only by the element characters, i.e., the characters that eat energy and those that store energy. The battery is an agent of the outside world. This agent imposes a voltage differential, i.e., an imposed *effort*, on the circuit that causes current to *flow*. As in Chapter 2, let's choose a node in the circuit between the resistor and the storage device, the inductor. Since the only charge storage element is the inductor (and this character is outside of the RVE), the current flowing into the node must exactly balance the current flowing out.

This concept must now be rendered mathematically. The current that passed through the energy dissipater must equal that flowing into the kinetic energy storage element.

$$\frac{V_O(t) - V_1(t)}{R} = i_R(t) = i_L(t).$$

We now introduce the relation corresponding to the storage of kinetic energy by virtue of flow:

$$V_1(t) - V_{REF} = V_1(t) = L\frac{di(t)}{dt}.$$

Substituting for $V_1(t)$ using the dissipation element relation:

$$V_1(t) - V_{REF} = V_1(t) = L\frac{di(t)}{dt} = V_O(t) - i(t)R$$

$$\frac{L}{R}\frac{di(t)}{dt} + i(t) = \frac{1}{R}V_O(t).$$

This differential equation governs the circuit current, i. It is important to note at this point that all the resistance, inductance, and interior system current again lie mathematically on one side of the equation while the driver from the outside world, the battery voltage, lies on the other side of the equation. The left side contains all system parameters and quantities while the right-hand side represents a forcing function that drives the flow.

$$\frac{L}{R}\frac{di}{dt} + i = \frac{1}{R}V_O(t)$$

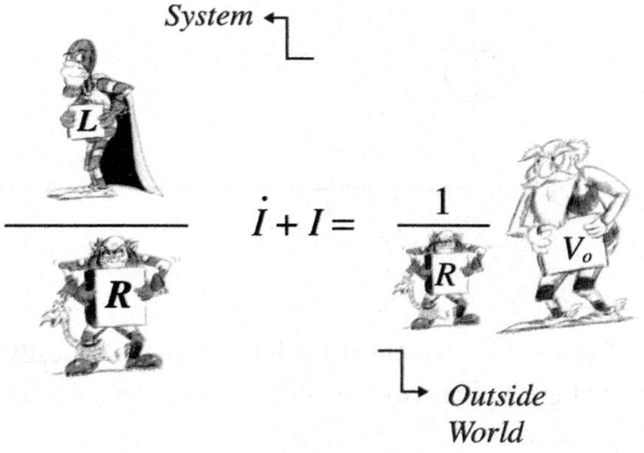

$$i + I = \frac{1}{R}V_o$$

System ←

Outside World

Figure 3.8: The electrical cast of characters playing out a series RL circuit.

> **Of Special Note**
>
> *This will be a constant theme in our development. The movie characters and their behavior live "on the left" while the circumstances presented to them by the outside world will lie "on the right."*

Here, the mathematical term on the right-hand side, $\frac{1}{R}V_O(t)$, is a flow-like external signal input to the system supplied by the battery.

3.4.3 A GENERALIZED MATHEMATICAL FORM FOR THE SINGLE STORAGE ELEMENT SCRIPT

If we observe the general nature of the governing differential equations for both the RC and RL circuits, there is a distinct one-to-one correspondence of terms. Single storage element scripts are characterized by 1st order ordinary differential equations in time. We further see that these equations can be cast in a form wherein:

(a) Either the effort or flow variable appears isolated with a coefficient of unity and

(b) The coefficient of the effort or flow derivative term (RC or L/R) has units of time

We generalize the governing ODE for any single storage element script or 1st order system as:

$$\tau \frac{d\psi}{dt} + \psi = \Psi_O(t) = G * I(t)$$

where τ is the system *time constant* and ψ is either an effort or flow variable in the system. Here, we have already seen cases where $\tau = RC$ and $\tau = L/R$. In general, the time constant will be some function of the system element parameters, $\tau = f(L, C, R)$. The forcing function will be some normalized form of the actual physical input excitation that renders an equivalent effort or flow driving function. The generalized forcing function is often represented as the actual physically imposed agent of excitation scaled by a factor called the static gain, G, where $\Psi_O(t) = G * I(t)$.

For the series RC circuit, we have the relations summarized in Table 3.3.

Table 3.3: Parts of 1st order governing differential equations for a series RC circuit

Response Variable	Capacitor Voltage, $V_1(t)$	Circuit Current, $i(t)$
System Parameter	$\tau = RC$	$\tau = RC$
External Excitation	$\Psi_o(t) = V_o(t)$	$\Psi_o(t) = C\dot{V}_o(t)$
	$G = 1; I(t) = V_o(t)$	$G = CD^1; I(t) = V_o(t)$

[1] Here, we use the differential operator where, by example, for an arbitrary variable, p: $Dp \equiv \dot{p} \equiv \frac{dp}{dt}$

While for the series RL circuit, we arrive at the results summarized in Table 3.4.

Table 3.4: Parts of 1st order governing differential equations for a series RL circuit

Response Variable	Inductor Voltage, $V_1(t)$	Circuit Current, $i(t)$
System Parameter	$\tau = L/R$	$\tau = L/R$
External Excitation	$??^2$	$\Psi_o(t) = \frac{V_o(t)}{R}$
	$G = ?? ; I(t) = ??$	$G = 1/R; I(t) = V_o(t)$

[2] These quantities will be asked of the reader in the Chapter Activities following this chapter.

One of the most powerful aspects of the analogical approach is that when systems behave linearly, the solutions to any equation expressed in this generalized form are essentially equivalent, i.e., ALL linear first order systems share inherent and important common characteristics in their system response to input or excitation from "the outside world." We will examine these common characteristics in detail when we address time domain solutions in Chapter 5.

Of Special Note
Universal Truths for 1st Order Systems

(a) They are comprised of system elements (or characters) that store ONLY ONE form of energy, *either* potential *or* kinetic forms of energy, *but not both.*

(b) Their behavior is characterized by a single system parameter called the system time constant, τ, where

(c) $\tau = f(R, C)$ *or* $\tau = g(L, R)$

3.5   MULTIPLE STORAGE ELEMENT SCRIPTS

The story changes when a system can store energy in more than one form. A more general circuit would be able to store electrical energy in both potential and kinetic forms as well as dissipate energy. The multiple storage element character script involves a capacitor, inductor, and resistor.

3.5.1   SERIES RLC CIRCUITS

Such a system is characterized by system capacitance, inductance, and electrical resistance. Consider a circuit where these elements are connected in series.

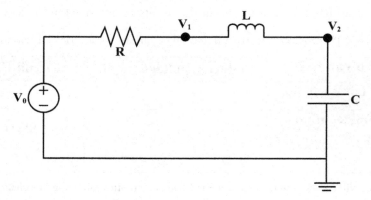

Figure 3.9: A series electrical RLC series circuit. Upon applying the battery to the circuit, current is driven in a clockwise sense around the circuit.

In this script, the battery hurls charge at the resistor which "eats" a portion, allowing some residue of the charge through to the inductor and capacitor. Charge build-up across the capacitor provides a voltage drop whose time rate of change corresponds to a time rate of charge across the capacitor plates. What occurs physically is that charge accumulates on one side of the capacitor. If the rate is sufficient to cause a rate of change of the voltage drop across the capacitor, charge at the other plate changes over time. Mathematically, at least, this dictates a current or effective movement of charge. This charge then "gets a boost from the battery" and starts the process all over again.

Focus on the voltage drop across the capacitor as the relevant system variable whose response we desire. Writing a current balance on node 2:

$$\dot{Q}_{IN} - \dot{Q}_{OUT} = \dot{Q}_{STORED}$$
$$i_L(t) = i_C(t)$$
$$\frac{V_1(t) - V_2(t)}{LD} = C\frac{d(V_2(t) - V_{REF})}{dt}$$
$$LC\ddot{V}_2(t) + V_2(t) = V_1(t).$$

But we do not know the other system voltage, $V_1(t)$. This is because there is now more than one way to store energy! Therefore, we must investigate a second current (or charge) balance at node 1.

$$\dot{Q}_{IN} - \dot{Q}_{OUT} = \dot{Q}_{STORED}$$
$$i_R(t) = i_L(t).$$

From the relation governing effort and flow through the resistor:

$$V_1(t) = V_O(t) - Ri_R(t)$$

and

$$i_R(t) = i_L(t)$$

but

$$i_L(t) = i_C(t) = C\dot{V}_2(t)$$

so

$$V_1(t) = V_O(t) - RC\dot{V}_2(t).$$

Substituting this into the relation obtained at the first node and rearranging terms:

$$LC\ddot{V}_2(t) + RC\dot{V}_2(t) + V_2(t) = V_O(t).$$

Once again, all the system parameters (R, L, C) and a voltage internal to the system, $V_2(t)$, are all on one side of the equation while the excitation "force" or effort supplying charge to the system "from the outside world" appears on the other side of the equation.

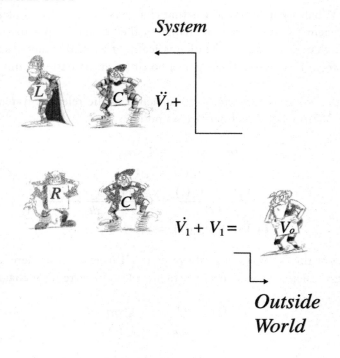

Figure 3.10: A series RLC electrical circuit character equation.

3.5.2   PARALLEL RLC CIRCUITS

One may also investigate a branched loop over which the charge will "choose the path of least resistance," or, more properly, the path of least impedance. The impedance is nothing more than a dynamic resistance. Using the definition of resistance as the ratio of effort/flow:

For the inductor:
$$\Delta V = LDi(t)$$
$$R_{DYNAMIC}^{INDUCTOR} = LD$$

For the capacitor:
$$CD\Delta V = i$$
$$\Delta V(t) = i(t)/CD$$
$$R_{DYNAMIC}^{CAPACITOR} = 1/CD$$

where, again, we are using the differential operator, $D(\bullet) = \dfrac{d(\bullet)}{dt}$.

Let's next consider a parallel RLC circuit in Figure 3.11.

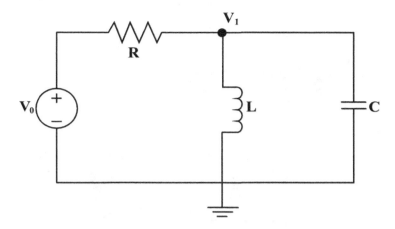

Figure 3.11: A parallel electrical RLC series circuit. Upon applying the battery to the circuit, current is driven in a clockwise sense around the circuit, but must now "choose the path of least impedance" at the branch point.

Performing a charge balance over the system at internal node 1:

$$\dot{Q}_{IN} - \dot{Q}_{OUT} = \dot{Q}_{STORED}$$
$$i_R(t) = i_L(t) + i_C(t)$$
$$\frac{V_O(t) - V_1(t)}{R} = \frac{V_1(t)}{LD} + C\dot{V}_1(t).$$

Applying the operator LD to both sides of the equation:

$$LC\ddot{V}_1(t) + \frac{L}{R}\dot{V}_1(t) + V_1(t) = \frac{L}{R}\dot{V}_O(t).$$

All the system parameters (R, L, C) and a voltage internal to the system, $V_1(t)$, are all on one side of the equation while the excitation "force" supplying charge to the system "from the outside world" appears on the other side of the equation.

In this script, the battery hurls charge at the resistor (Evil Dr. Friction). Evil Dr. Friction eats some charge allowing less out which then is stored in the system inductor (in the form of electrical kinetic energy) and/or the capacitor (in the form of electrical potential energy). How much is stored in each of these storage elements depends on their impedance or instantaneous (dynamic) electrical resistance with more energy being stored in the path with least impedance.

When the storage characters dominate over friction, they will pass energy back and forth with friction eating away at each transfer. In Figure 3.13, a system imparted with potential energy

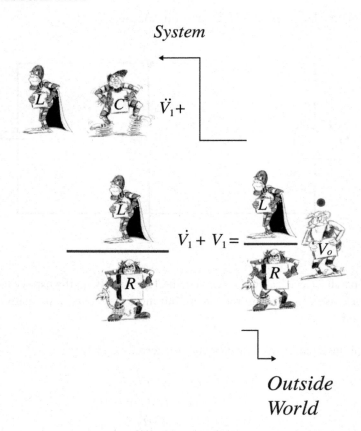

Figure 3.12: A parallel electrical RLC series circuit character equation.

(A) will pass it on to Captain Kinetic Energy. Dissipation is eating energy during this transfer as evidence by Evil Dr. Friction fighting Captain Kinetic Energy (B). Dissipation continues to degrade the energy cache during each subsequent exchange back to Captain Potential Energy (C) and back to Captain Kinetic Energy (D) until all the electrical energy has been consumed. In the case where an input signal delivers energy continually to the system, eventually the amount stored in potential and kinetic forms reaches a steady state while the energy losses continue to accrue with time.

Figure 3.13: A second order system with dissipation results in energy being "consumed" within each exchange from kinetic to potential and back to kinetic.

3.5.3   IDEALIZED LC CIRCUITS

Consider the first series RLC circuit. In the limit as the resistance vanishes, the differential equation for the capacitor voltage becomes:

$$LC\ddot{V}_2(t) + V_2(t) = V_O(t).$$

In this script, the battery provides a voltage or effort difference that drives charge at the inductor. A charge difference causes a rate of change of current passing through the inductor. This process creates kinetic energy that is present in the system owing to the presence of the inductor. Because charge must be conserved, the rate of change of current in the inductor results in a charge difference across the capacitor that varies in time, i.e., changes in stored potential energy in the capacitor. Charge differences that change in time across the plates of the capacitor result in a flow of charge. This charge flows back to the battery for "a boost" from the "outside world." The electrical energy is simply transferred from kinetic to potential and back with no dissipation ad infinitum. We will "see" this behavior in the structure of the mathematical solutions described in Chapter 5. This system is a simple frictionless harmonic oscillator where the harmonic response occurs in the system voltage or effort variable as well as the current or flow variable. This is analogous to motion in a simple, frictionless pendulum where a similar simple harmonic motion results for the angular velocity and position (flow variables). We can also show that harmonic variation also occurs in the component of the gravitational force that produces the internal torque driving the system back again.

Figure 3.14: A second order system without dissipation results in energy simply being transferred between potential and kinetic forms, but otherwise being conserved in total. The simple pendulum is an analog to the LC circuit in the absence of any electrical resistance.

3.5.4   A GENERALIZED MATHEMATICAL FORM FOR THE DUAL STORAGE ELEMENT SCRIPT

If we examine the governing ordinary differential equations for the system voltages in Sections 3.5.1, 3.5.2, and 3.5.3, we see that dual storage element scripts are always characterized by 2nd order ordinary differential equations in time. We can further see that the resulting governing differential equations can be cast in a form where:

(a) the effort or flow variable appears isolated with a coefficient of unity on "the system side" of the ODE,

(b) the coefficient of the effort or flow derivative term (RC or L/R) has units of time, and

(c) the coefficient of the effort or flow second derivative term (LC) has units of $[T]^2$

One can then generalize the governing ODE for any dual storage element script or 2nd order system as:

$$\frac{1}{\omega_N^2}\frac{d\psi}{dt} + \frac{2\zeta}{\omega_N}\frac{d\psi}{dt} + \psi = \Psi_O(t) = G * I(t),$$

where ω_N is the system natural frequency, ζ is the dimensionless system damping ratio, and ψ is either an effort or flow variable in the system. Here, we have already seen a similar situation when the equation is 1st order. In this case, we saw the $\tau = RC$ or $\tau = L/R$. For 2nd order systems, there are two system parameters: the natural frequency and damping ratio will be functions of the system element parameters, $\{\omega_N, \zeta\} = f(L, C, R)$. The forcing function will be some normalized form of the actual physical input excitation that renders an equivalent effort or flow driving function. The generalized forcing function is often represented as the actual physically imposed agent of excitation scaled by a factor called the static gain, G, where $\Psi_O(t) = G * I(t)$. For the second order RLC circuits, the results obtained are summarized in Table 3.5.

Table 3.5: Parts of 2nd order governing differential equations for series and parallel RLC circuits

RLC Circuits	Series Circuit	Parallel Circuit
Response Variable	Capacitor Voltage, $V_2(t)$	Capacitor/Inductor Voltage, $V_1(t)$
System Parameter	$\omega_N = \dfrac{1}{\sqrt{LC}}$; $\zeta = \dfrac{R}{2}\sqrt{\dfrac{C}{L}}$	$\omega_N = \dfrac{1}{\sqrt{LC}}$; $\zeta = \dfrac{1}{2R}\sqrt{\dfrac{L}{C}}$
External Excitation	$\Psi_o(t) = V_o(t)$	$\Psi_o(t) = \dfrac{L}{R}\dot{V}_o(t)$
	$G = 1; I(t) = V_o(t)$	$G = LD/R;\ I(t) = V_o(t)$

Of Special Note
   Universal Truths for 2nd Order Systems

(a) They are comprised of system elements (or characters) that store BOTH potential AND kinetic forms of energy

(b) Their behavior is characterized by a pair of system parameters, $\{\omega_N, \zeta\}$, where

(c) $\omega_N = f(L, C)$ *and* $\zeta = g(L, C, R)$

One of the most powerful aspects of the analogical approach is that when systems behave linearly, the solutions to any equation expressed in this generalized form are essentially equivalent, i.e., ALL linear second order systems share inherent and important common characteristics in their system response to input or excitation from "the outside world." We will examine these common characteristics in detail when we address time domain solutions in Chapter 5.

3.6 CHAPTER ACTIVITIES

Problem 1 Perform a charge balance over an appropriate circuit node in the series RC circuit in Figure 2.2 to derive a governing differential equation for the circuit's current (flow variable) instead of the circuit voltage drop across the capacitor (effort variable).

Problem 2 Perform a charge balance over an appropriate circuit node in the RL circuit in Figure 3.7 to derive a governing differential equation for the circuit's voltage drop (effort variable) across the inductor instead of the circuit current (flow variable). Fill in the missing normalized excitation signal input in Table 3.3.

Problem 3 Recast the nodal balance over the representative circuit nodes in the series RLC circuit to derive a governing differential equation for the circuit's current (flow variable) instead of the system voltage drop across the capacitor (effort variable).

Problem 4 Recast the nodal balance over the representative circuit nodes in the parallel RLC circuit in to derive a governing differential equation for the circuit's current (flow variable) instead of the system voltage drop across the inductor/capacitor pair (effort variable).

Problem 5 Consider the series circuit for which the capacitor and resistor are swapped resulting in a series CR circuit shown here:

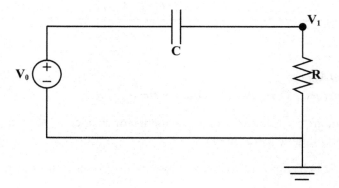

Perform a charge balance at an appropriate node and derive the differential equation governing the voltage drop across the resistor.

Problem 6 An actual inductor will often contain non-negligible resistance because it is a long coiled piece of wire. For this more realistic version of the parallel RLC circuit shown here:

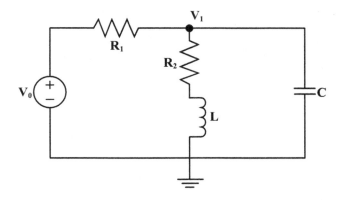

perform a charge balance at an appropriate node and re-derive the governing differential equation for the voltage drop across the capacitor.

Problem 7 Consider the circuit shown with parallel system capacitors. At $t = 0$, a step voltage, V_0, is applied to the circuit by connecting it suddenly across a battery:

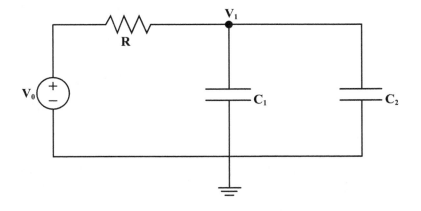

$$e_0 = 12 \, \text{V}$$
$$i_R(t = 0) = 40 \, \textit{milliamps}.$$

(a) On the circuit diagram label the relevant nodes and apply the necessary conservation principles to derive the differential equation governing the response of the voltage drop across the pair of capacitors in the circuit.

   (b) What order is the equation? Use the potential energy storage system element equation to find the relevant initial condition or initial conditions for the system effort variable.

   (c) Compare the governing equation with that from the simple series RC circuit in Figure 2.2. What conclusion can you draw about the effective capacitance of a pair of capacitors in parallel?

CHAPTER 4

The Mechanical Cast

A body perseveres in its state of being at rest or of moving uniformly straight forward, except insofar as it is compelled to change its state by forces impressed.

Sir Isaac Newton

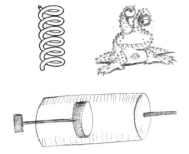

This transfer of knowledge from one branch of science, electrical network theory, to another branch of science dealing with mechanical structures is one of a long line of such interchanges (that are) made possible by fundamental analogies which rest finally on that fact that electrical and mechanical motions satisfy the same type of differential equations. Since these interchanges have been going on for hundreds of years, it seems worthwhile to examine their foundation and development.

W.P. Mason
"Electrical and Mechanical Analogies"
Bell System Technical Journal

In Chapter 1, we examined the concepts of effort and flow which continue to guide and build our analogy between different disciplines. Recall that we posited that force acts as an effort to cause motion. Thereby, the flow variable can be represented by either the displacement or velocity variable depending on whether one wishes to choose the motion variable or its rate of change in time as the pertinent flow variable. With these choices made, rectilinear forces that act on a mass cause changes to the directional momentum of the mass. As per Newton, the net force acting on a mass equals the net change in momentum. In the absence of a net force, the linear momentum of a mass or particle is conserved.

4.1   EFFORT AND FLOW VARIABLES

If you push inertia, it will flow. The flow of mass is, by our definition, velocity. How you push a mass is by creating a force differential or a net force across the mass. This is clearly evidenced in a free body diagram. According to Newton's second law of motion, the net force applied to an inertial element results in a time rate of change in its linear momentum. This is a statement of the balance of linear momentum, as summarized in Table 4.1.

Table 4.1: Effort, flow, and conserved quantities for translational mechanical systems

Conserved Quantity			Units	Symbol
Linear momentum			kg-m/s	p
Variable			Units	
Effort	Force		N ; lb	F
Flow	Velocity		m/s ; ft/s	υ

4.2   STORAGE ELEMENTS

It is now time to identify the mechanical cast that will play the roles of energy storage and dissipation in mechanical systems. Typically, the motion can be separated into translational and rotational components. These can be analyzed separately in linear systems.

4.2.1   POTENTIAL ENERGY STORAGE CHARACTER

The mechanical cast member who plays the role of Captain Potential Energy is that device that stores a force internally that may, at some later time, be released to perform useful mechanical work on the system. This potential energy storage character is played by the simple spring.

Figure 4.1: The mechanical potential energy storage character is played by the spring. It embodies the mechanical capacitance of the system.

The governing mathematical expression of the storage by virtue of effort is

$$FLOW = C_{MECH}\frac{d(EFFORT)}{dt}$$

$$\upsilon = C_{MECH}\frac{d(F_{NET})}{dt}.$$

Integrating both sides over time results in an expression for the mechanical analog to an electrical system's capacitance

$$\frac{1}{C_{MECH}}\int \upsilon(t)dt = F_{NET} = kx$$

$$C_{MECH} \equiv \frac{1}{k}.$$

4.2.2   KINETIC ENERGY STORAGE CHARACTER

Using the rate form, we address the flow rate of position or velocity. The mechanical cast member who plays the role of Captain Kinetic Energy is that device that stores energy by virtue of its flow or velocity. The mechanical actor who stores kinetic energy by virtue of velocity is the system's inertia.

Figure 4.2: The mechanical kinetic energy storage character is played by the system's inertia. Inertia is embodied in a system's mass.

The governing mathematical expression of the storage by virtue of flow is

$$EFFORT = L\frac{d(FLOW)}{dt}$$

$$F_{NET} = L_{MECH}\frac{d\upsilon}{dt} = ma.$$

Since this relation gives us Newton's second law of motion, one concludes that the mechanical analog to an electrical system's inductance is the inertia or mass of the mechanical system

$$L_{MECH} \equiv m.$$

4.3   DISSIPATIVE ELEMENTS

The role of the Evil Dr. Friction in a mechanical script is played by the physical presence of friction. Friction, in a sense, eats energy, reducing the amount available to produce motion.

Figure 4.3: The mechanical energy character that dissipates energy is played by any form of friction. Here the friction acts physically along the surface of some inertia with the floor on which it is sliding. Father Force performs work on the mass which it can store as kinetic energy of motion thwarted by the Evil Dr. Friction who eats a portion of the input work done by Father Force.

The governing mathematical expression of the dissipation is always algebraic rather than differential. If we consider the source of the friction to be viscous friction as, say, would result from a thin layer of viscous oil between the box and the floor. Alternatively, the same force would result in a mechanical damper in which the same shear force is developed in a cylindrical dashpot. The viscous force resisting the relative motion of the ends of the dashpot is proportional to the relative velocity

$$F_{NET} = bv.$$

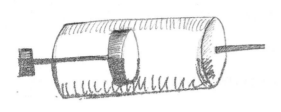

Figure 4.4: The friction force is modeled by the net force across a mechanical dashpot; this viscous force is proportional to the relative velocity. For many applications, a viscous representation of friction may suffice.

The resistive effort flow relation is also algebraic

$$EFFORT = R * FLOW$$
$$F_{NET} = R_{MECH} v.$$

This relation dictates that the mechanical analog to the electrical system's resistance is the viscous friction or damping coefficient, b

$$R_{MECH} \equiv b.$$

Alternatively, the friction could result from other physical sources such as dry friction, often termed Coulomb friction. Many systems, however, have friction forces that may be described as viscous-like in nature, enough so that the algebraic relation between the dissipation and flow holds. A summary of the mechanical cast and the roles they play is given in Figure 4.5. A list of corresponding system element equations is given in Table 4.2.

4.4 SINGLE STORAGE ELEMENT SCRIPTS

4.4.1 SPRING-DAMPER SYSTEMS

An idealized case often studied is that of the mass-less spring-damper system. This represents the bound on behavior of a system with negligible inertia that is dominated by its elasticity and friction. In the case of the spring-damper system, mechanical energy is lost through the damper while the residue is stored by virtue of a net force inside the spring.

Translational Mechanical Systems

Conservation of LINEAR MOMENTUM

EFFORT Variable = Force
FLOW Variable = Velocity (displacement)

Potential Energy Storage Actor

Linear Spring

Kinetic Energy Storage Actor

Inertia (mass)

Energy Dissipation Actor

Damper

Figure 4.5: The mechanical cast of characters.

Table 4.2: Relevant system element relations for translational mechanical systems

Field	Effort Variable	Flow Variable
Mechanical	Force	Velocity
Relation	Form	Analogy
Dissipative Material Property Law	Effort = Resistance x Flow $$F = b\left(\dot{x}_2 - \dot{x}_1\right)$$	Resistance = Friction; Damping coefficient, b
Energy Storage in Effort Variable	Flow = Capacitance x d(Effort)/dt $$v = \frac{1}{k}\frac{dF}{dt}$$	Capacitance = $$\frac{1}{k} = \frac{1}{stiffness}$$
Energy Storage in Flow Variable	Effort = Inductance x d(Flow)/dt $$F = m\frac{dv}{dt}$$	Inductance = Mass/Inertia, m

Figure 4.6: An idealized mass-less spring-damper system under the influence of an externally applied force, $F(t)$.

In order to balance linear momentum of the inertia-less plate, we perform a force balance on a representative piece of the system, i.e., the plate. For mechanical systems, this part of the system is that on which all forces act, the mass. The result is a free body diagram (FBD).

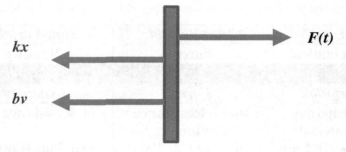

Figure 4.7: Free body diagram (FBD) for a mass-less spring-damper system.

Of Special Note

> *Free body diagrams are the representative volume elements (RVE) for all mechanical systems.*

Summing all forces and assuming the mass is constant:

$$\dot{Q}_{IN} - \dot{Q}_{OUT} = \dot{Q}_{STORED}$$

$$F_O(t) - kx - b\dot{x} = \frac{dp}{dt} = m\frac{dv}{dt} = 0.$$

Rearranging terms results in the differential equation governing position of the plate

$$b\dot{x} + kx = F_O(t)$$

$$\frac{b}{k}\dot{x} + x = \frac{1}{k}F_O(t)$$

This differential equation is linear and first order. Appealing to our analogy with electrical systems:

$$\frac{b}{k}\dot{x} + x = b\frac{1}{k}\dot{x} + x = R_{MECH}C_{MECH}\dot{x} + x = \frac{1}{k}F_O(t) = C_{MECH}F_O(t)$$

where

$$R_{MECH} = b$$

$$C_{MECH} = 1/k.$$

A similar system character equation results analogous to the electrical RC circuit in Figure 4.8.

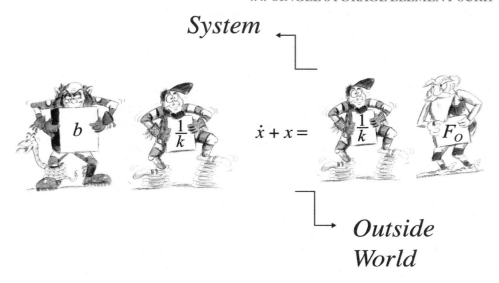

$$\dot{x} + x =$$

Figure 4.8: The mass-less spring-damper mechanical system is a purely mechanical analog to the series RC circuit as evidenced by the character equation.

Unlike electrical systems, mechanical systems' differential equations are not typically "mathematically versatile" in that they will almost exclusively appear with the flow variable as the dependent variable. The equation governing the plate displacement could be re-cast in terms of the reaction force necessary to cause a given displacement, but this is often relegated to post-processing the displacement solution, i.e., one typically does NOT see differential equations for the force stored in or transmitted by the spring where force is solved as the primary variable. In most, if not all, mechanical systems, the primary solution variable is the flow variable.

4.4.2 MASS-DAMPER SYSTEMS

What if we wanted to examine a system that stored its energy solely in kinetic form? The alternative two-character script with a single energy storage character would involve Captain Kinetic Energy battling the Evil Dr. Friction! Consider a parachutist diving out of an airplane and suddenly pulling their chute cord. They're subject to a step input force from gravity. Father Force is instantaneously pulling them toward the Earth, as illustrated in Figure 4.9. The Evil Dr. Friction is also pushing back the parachute with a drag force due to the air in the parachute. In this case, one may argue that friction is not so evil as it fights gravity. But if we view motion of the mass as giving the system kinetic energy, then friction continues to eat that energy away from the diver. In this case, friction happens to be our friend (if we desire a safe landing), but it is still the enemy of speed. Aerodynamic drag forces are always functions of the skydiver's downward velocity. For simplicity, let's say the drag force is linearly proportional to the velocity. We begin modeling this system with a free body diagram shown in Figure 4.9.

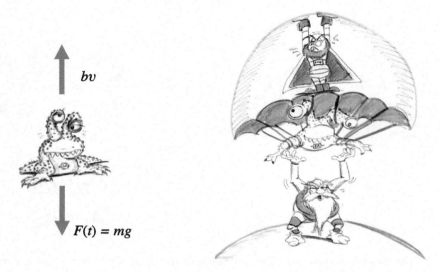

Figure 4.9: Upon diving from an airplane, a skydiver experiences a sudden step input force exerted by gravity. The parachute provides a velocity-dependent drag force opposing the gravitational force. The net force results in the diver's acceleration.

At this point, we should point out that the *system* is defined only by the element characters. Father Force is Planet Earth, providing a driving effort that is an energy supply to the system from the outside world. Normally this energy would turn entirely into kinetic energy of the skydiver with potentially fatal results. But the Evil Dr. Friction consumes part of the energy. The rest is stored by way of velocity in the mass of the diver by Captain Kinetic Energy.

Each system element character exhibits its own characteristic effort-flow equation. So by balancing linear momentum:

$$\dot{Q}_{IN} - \dot{Q}_{OUT} = \dot{Q}_{STORED}$$
$$mg - b\dot{x}(t) = \frac{dp(t)}{dt} = m\frac{dv(t)}{dt} = m\ddot{x}(t)$$
$$m\ddot{x}(t) + b\dot{x}(t) = mg.$$

Recognizing that this differential equation is actually first order in velocity

$$\frac{m}{b}\ddot{x}(t) + \dot{x}(t) = \frac{1}{b}mg = v_{TERMINAL}$$
$$\frac{m}{b}\dot{v}(t) + v(t) = \frac{1}{b}mg = v_{TERMINAL}.$$

This differential equation governs the system flow variable or velocity of the mass, υ. The left side contains all system parameters and variables while the right-hand side represents a scaled forcing function that drives the flow to its steady-state value, the so-called terminal velocity! So we don't need to solve the equation to see where it's heading. The equation can be written in an effectively identical form to that governing the electrical RL circuit in Figure 3.10.

$$\frac{m}{b}\frac{d\upsilon(t)}{dt} + \upsilon(t) = \frac{1}{b}F_O(t).$$

This equation, in fact, takes on a form identical to the RL circuit when the analogous mechanical parameters are introduced

$$\frac{L_{MECH}}{R_{MECH}}\frac{d\,(FLOW)}{dt} + FLOW = \frac{1}{R_{MECH}}F_O(t)$$
$$= \frac{EXTERNAL\,EFFORT}{R_{MECH}} = FLOW_{SS}.$$

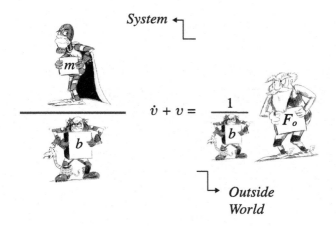

Figure 4.10: The skydiving mass-damper mechanical system is a purely mechanical analog to the series RL circuit as evidenced by the character equation.

4.4.3   A GENERALIZED MATHEMATICAL FORM FOR THE SINGLE STORAGE ELEMENT SCRIPT

If we observe the governing differential equations for both the spring-damper and mass-damper mechanical systems, we see that single storage element scripts are characterized by the same 1st order ordinary differential equations in time as 1st order electrical systems:

$$\tau\frac{d\psi(t)}{dt} + \psi(t) = \Psi_O(t) = G * I(t)$$

where τ is the system time constant and ψ is either an effort or flow variable in the system. The forcing function will be some normalized form of the actual physical input excitation that renders an equivalent effort or flow driving function.

For the parallel spring-damper system, the first order time constant and signal excitation are summarized in Table 4.3.

Table 4.3: Parts of 1st order governing differential equations for a parallel $k–b$ system

Response Variable	Platform Position, $x(t)$
System Parameter	$\tau = \dfrac{b}{k} = R_{MECH} C_{MECH}$
External Excitation	$\Psi_o(t) = \dfrac{F_0(t)}{k}$
	$G = 1/k \; ; \; I(t) = F_O(t)$

While for the mass-damper system, an analogous set of relations is summarized in Table 4.4.

Table 4.4: Parts of 1st order governing differential equations for a mass-damper system

Response Variable	Platform Position, $x(t)$
System Parameter	$\tau = \dfrac{m}{b} = \dfrac{L_{MECH}}{R_{MECH}}$
External Excitation	$\Psi_o(t) = \dfrac{F_0(t)}{b} = \dfrac{mg}{b}$
	$G = 1/b \; ; \; I(t) = F_O(t) = mg$

One of the most powerful aspects of the analogical approach is that when systems behave linearly, the solutions to any equation in this generalized form are essentially equivalent, i.e., ALL linear first order systems share inherent characteristics in their system response to input or excitation from "the outside world."

4.5   MULTIPLE STORAGE ELEMENT SCRIPTS

4.5.1   THE CLASSICAL MASS-SPRING-DAMPER SYSTEM

Introducing non-negligible inertia to the platform in Section 4.3, the two-storage-element-character script now has the ability to store kinetic as well as potential energy as depicted in Figure 4.11.

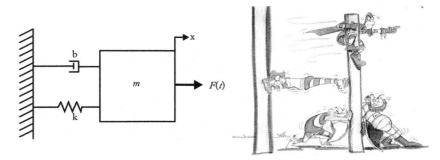

Figure 4.11: A parallel m–k–b mechanical system. Upon applying the external force to the inertial element, flow or motion of the mass is driven.

Writing a linear momentum balance on the mass:

$$\dot{Q}_{IN} - \dot{Q}_{OUT} = \dot{Q}_{STORED}$$

$$F_O(t) - kx(t) - b\dot{x}(t) = \frac{dp(t)}{dt} = m\frac{dv(t)}{dt} = m\ddot{x}(t).$$

Rearranging terms and normalizing the equation

$$m\ddot{x}(t) + b\dot{x}(t) + kx(t) = F_O(t).$$

After scaling the entire equation by the stiffness to normalize the flow variable term

$$\frac{m}{k}\ddot{x}(t) + \frac{b}{k}\dot{x}(t) + x(t) = \frac{1}{k}F_O(t).$$

Using the mechanical analogs for the electrical system element parameters

$$L_{MECH}C_{MECH}\ddot{x}(t) + R_{MECH}C_{MECH}\dot{x}(t) + x(t) = \frac{1}{k}F_O(t)$$

where

$$L_{MECH} = m$$
$$C_{MECH} = 1/k$$
$$R_{MECH} = b.$$

In this script, the external excitation provided by Father Force translates into a change of momentum of the inertia in the system. The resistance acting against the mass in the form of the Evil Dr. Friction eats some work performed on the block with the residual work being stored as potential energy that stretches the spring and kinetic energy stored by virtue of the velocity of the mass.

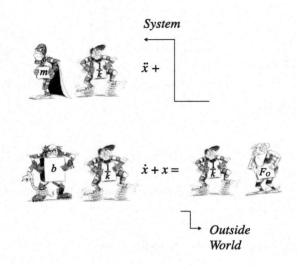

Figure 4.12: A classical mass-spring-damper system is equivalent to the series RLC circuit when placed in the form of a character equation.

4.5.2   IDEALIZED MASS-SPRING SYSTEMS

An idealized case can be illustrated when the resistance becomes "infinitesimally small" and ALL of the energy input by the driving force is stored as potential energy in the spring and kinetic energy in the mass. This is the idealized case of a system without losses. In this case:

$$m\frac{1}{k}\ddot{x}(t) + x(t) = \frac{1}{k}F_O(t)$$
$$L_{MECH}C_{MECH}\ddot{x} + x = C_{MECH}F_O(t).$$

In this script, the external force drives a momentum change in the mass which is slowed down by the spring opposing its motion. As the kinetic energy imparted to the mass by the force is reduced, an equivalent amount of potential energy is stored in the spring. The mechanical energy is simply transferred from kinetic to potential and back with no dissipation ad infinitum. In this sense, Captain Potential Energy and Captain Kinetic Energy "have a catch" with a ball of energy while the Evil Dr. Friction gets none. This is the translational mechanical analog to the equivalent LC electrical circuit. Recall, we made an appeal to our intuition about a simple frictionless pendulum system at the end of Chapter 3. The frictionless pendulum is a rotational mechanical analog of the simple mass-spring harmonic oscillator discussed here and the idealized LC circuit of Section 3.5.3.

It is now a natural excursion to relate the effort-flow story for such rotational mechanical systems in Section 4.6.

Figure 4.13: The simple frictionless pendulum is an analog system to the LC circuit in the absence of any electrical resistance.

Table 4.5: Parts of a 2nd order governing differential equation for a classical m–k–b system

Mechanical Systems	Classical m-k-b System
Response Variable	**Position,** $x(t)$
System Parameter	$\omega_N = \sqrt{\dfrac{k}{m}} \;\; ; \;\; \zeta = \dfrac{b}{2\sqrt{km}}$
External Excitation	$\Psi_O(t) = F_O(t)\Big/ k$
	$G = 1/k \; ; \; I(t) = F_O(t)$

4.5.3 A GENERALIZED MATHEMATICAL FORM FOR THE DUAL STORAGE ELEMENT SCRIPT

All mechanical scripts in which two distinct energy storage characters appear are always characterized by the same 2nd order ordinary differential equations in time as electrical 2nd order systems:

$$\frac{1}{\omega_N^2}\frac{d\psi(t)}{dt} + \frac{2\zeta}{\omega_N}\frac{d\psi(t)}{dt} + \psi(t) = \Psi_O(t) = G * I(t)$$

where ω_N is the system natural frequency, ζ is the dimensionless system damping ratio, and $\{\omega_N, \zeta\} = f(L, C, R)$. The dependet variable, ψ, is either an effort or flow variable in the system. The forcing function will be some normalized form of the actual physical input excitation that renders an equivalent effort or flow driving function.

Table 4.5 summarizes the results for the second order mechanical systems discussed so far.

4.6   ROTATIONAL MECHANICAL SYSTEMS

One example is torque, moment of inertia, angular momentum, vs force, mass and momentum. The possible undistinguishability of translation and rotation would seem to indicate that they are really two guises for the same set of phenomena.

The Physics Stack Exchange

4.6.1   EFFORT AND FLOW VARIABLES

If you push a mass with a rectilinear or translational force, translational velocity will evolve over time as the mass accelerates. You push a mass by creating a difference in rectilinear force across the mass, i.e., applying a net force. It is precisely analogous to note that if you twist a rotational inertia, such as a massive disk, for example, it will develop an angular velocity. To do this, you need to apply a net rotational force or a net torque. All governing equations are based on writing mathematical statements of the conservation of angular momentum as summarized in Table 4.6.

Table 4.6: Effort, flow, and conserved quantities for rotational mechanical systems

Conserved Quantity		Units	Symbol
Angular momentum		$kg\text{-}m^2/s$	L
Variable		Units	
Effort	Torque	Nm ; ft-lb	T
Flow	Angular velocity	rad/s	ω

So, for rotational mechanical systems, one still draws an appropriately labeled free body diagram, only now one must sum the external torques and relate this to a net change in angular momentum according to Newton's laws. This is done in a manner strictly analogous with translational mechanical systems.

4.6.2   STORAGE ELEMENTS

The energy storage occurs through the same actors: springs for potential energy and masses for kinetic energy, but these must now be an angular or torsional spring, κ, and a measure of inertial resistance to angular motion or a mass moment of inertia, J.

Potential Energy Storage Character

The torsional potential energy storage devices store energy in the form of the effort variable or torque. The mechanical cast member who plays the role of Captain Potential Energy is that device that stores a torque internally that may, at some later time, be released to perform useful rotational form of mechanical work on the system. This potential energy storage character is played by the torsional spring.

Figure 4.14: The rotational mechanical potential energy storage character is played by the torsional spring. It embodies the rotational mechanical capacitance of the system.

The mathematical expression of the storage by virtue of effort is

$$FLOW = C_{ROT\_MECH} \frac{d(EFFORT)}{dt}$$

$$\omega(t) = C_{ROT\_MECH} \frac{dT_{NET}(t)}{dt}.$$

Integrating both sides results in an expression for a rotational mechanical analog to electrical capacitance

$$\frac{1}{C_{ROT\_MECH}} \int \omega(t)dt = T_{NET}(t) = \kappa\theta(t)$$

$$C_{ROT\_MECH} \equiv \frac{1}{\kappa}.$$

Kinetic Energy Storage Character

The mechanical cast member who plays the role of Captain Kinetic Energy is that device that stores energy internally by virtue of its rotational speed. This potential energy storage character is played by the rotational or mass moment of inertia.

You may recall from your undergraduate dynamics course that the rotational form of Newton's Second Law states that a net torque applied to a system is equal to the time rate of change

Figure 4.15: The rotational mechanical kinetic energy storage character is played by the system's rotary or mass moment of inertia. Inertia is embodied in a system's mass weighted by the square of its distance about the axis of rotation.

of the system's angular momentum, H. The mathematical expression of the storage by virtue of flow is

$$EFFORT = L\frac{d(FLOW)}{dt}$$

$$T_{NET}(t) = \frac{dH}{dt} = \frac{d(J\omega(t))}{dt} = L_{ROT\_MECH}\frac{d\omega(t)}{dt} = J\alpha(t) = J\ddot{\theta}(t).$$

Again, applying the effort-flow analogy, one observes that the mass moment of inertia is the mechanical analog of an electrical inductance

$$L_{ROT\_MECH} \equiv J \equiv \int r^2 dm.$$

4.6.3   DISSIPATIVE ELEMENTS

The role of the Evil Dr. Friction in our rotational mechanical script is played by any physical presence of friction about the axis of rotation. Let's consider the source of the friction to be viscous friction as would result from a thin layer of viscous oil between two rotational elements as in a sleeve bearing.

Figure 4.16: The friction force is modeled by the net torque across a mechanical cylindrical dashpot; this viscous force is proportional to the relative angular velocity. For many applications, a viscous representation of friction may suffice.

For which

$$EFFORT = R * FLOW$$
$$T_{NET} = R_{ROT\_MECH}\omega(t) = \beta\omega(t)$$
$$R_{ROT\_MECH} \equiv \beta$$

where β is a torsional damping coefficient relating the torque necessary to sustain an angular velocity differential across a rotational frictional element. A summary of the rotational mechanical cast and the roles they play is given in Figure 4.17. A list of corresponding system element equations is given in Table 4.7.

4.6.4 THE SIMPLE PENDULUM

Consider the swinging pendulum shown in Figure 4.18. If we perform an angular momentum balance about the pivot point:

$$\dot{Q}_{IN} - \dot{Q}_{OUT} = \dot{Q}_{STORED}$$
$$T_O(t) - mgL\sin\theta(t) - \beta\dot{\theta}(t) = \frac{dH(t)}{dt} = J\frac{d\omega(t)}{dt} = J\alpha(t) = J\ddot{\theta}(t).$$

Assuming small angles of rotation linearizes the system

$$\sin\theta(t) \approx \theta(t) \Rightarrow J\ddot{\theta}(t) + \beta\dot{\theta}(t) + mgL\theta(t) = T_O(t).$$

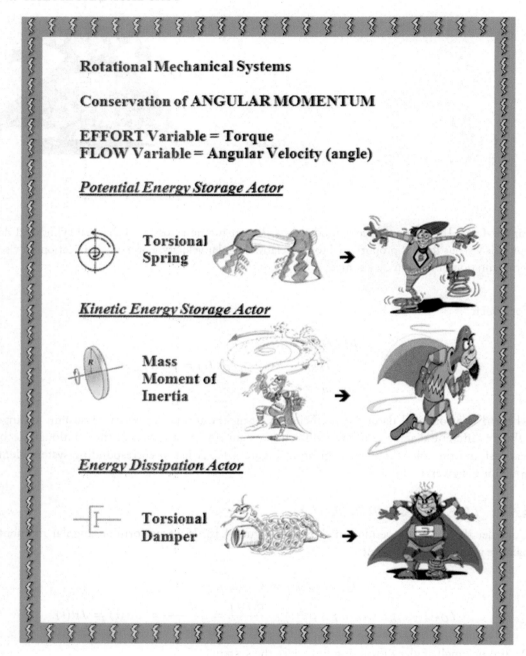

Rotational Mechanical Systems

Conservation of ANGULAR MOMENTUM

EFFORT Variable = Torque
FLOW Variable = Angular Velocity (angle)

Potential Energy Storage Actor

Torsional
Spring

Kinetic Energy Storage Actor

Mass
Moment of
Inertia

Energy Dissipation Actor

Torsional
Damper

Figure 4.17: The rotational mechanical cast of characters.

Table 4.7: Relevant system element relations for rotational mechanical systems

Field	Effort Variable	Flow Variable
Mechanical	Torque	Angular velocity
Relation	Form	Analogy
Dissipative Material Property Law	Effort = Resistance x Flow $T = \beta(\omega_2 - \omega_1)$	Resistance = Friction; Damping coefficient, β
Energy Storage in Effort Variable	Flow = Capacitance x d(Effort)/dt $\omega = \dfrac{1}{\kappa}\dfrac{dT}{dt}$	Capacitance = $\dfrac{1}{\kappa} = \dfrac{1}{stiffness}$
Energy Storage in Flow Variable	Effort = Inductance x d(Flow)/dt $T = J\dfrac{d\omega}{dt}$	Inductance = Rotary Mass/Inertia, J

Rearranging terms and normalizing by the effective torsional stiffness

$$\frac{J}{\kappa_{EFF}}\ddot{\theta} + \frac{\beta}{\kappa_{EFF}}\dot{\theta} + \theta = \frac{1}{\kappa_{EFF}}T_O(t)$$

$$\kappa_{EFF} = mgL.$$

All the system parameters (J, β, κ_{EFF}) and a flow variable internal to the system, θ, are all on one side of the equation while the excitation effort, now an applied torque, appears on the other side. If we scale the entire equation by the torsional stiffness to normalize the flow variable term

$$L_{ROT\_MECH}C_{ROT\_MECH}\ddot{\theta}(t) + R_{ROT\_MECH}C_{ROT\_MECH}\dot{\theta}(t) + \theta(t) = \frac{1}{\kappa_{EFF}}T_O(t).$$

Note that while you might not see a torsional spring here, there is one! By virtue of hanging from a cable of length, L, in a gravitational field, the mass may store maximum potential energy at the ends of each swing where height of the mass provides potential energy due to the work of the gravitational field. Gravity is our spring! Possible sources of damping are provided by air resistance during the swing and friction at the pivot. Rotational inertia is provided by the mass being lumped a finite distance from the pivot, the center of rotation. This is illustrated in Figure 4.18. Here, Father Force provides the effort or torque to drive the swinging angular motion. At the ends of the swing, all energy is potential in form. As the bob gains speed on the downswings, the system

Figure 4.18: THE rotational pendulum.

gains rotational kinetic energy at the expense of potential energy. All the while, the Evil Dr. Friction, acting in the air flowing around the bob and in resistance at the pivot, eats away at each exchange.

4.7 CHAPTER ACTIVITIES

Problem 1 It is somewhat intriguing and not often discussed what the mechanical system analogous to the parallel RLC circuit (discussed in Section 3.4.2) would be. Identify this mechanical system whose governing differential equation would be analogous with that obtained for the parallel RLC circuit. Draw the system elements and their relative connectivity and derive the governing differential equation.

Problem 2 Consider the plate damper, mechanical system from which the spring has been removed. The system is turned vertically and subject to a step input gravitational force as shown:

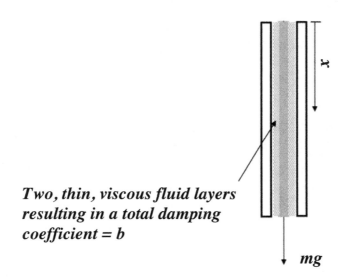

Two, thin, viscous fluid layers resulting in a total damping coefficient = b

If the mass is dropped from the position $x_O = 0\,m$ from rest, write the differential equation governing the system plate velocity. What order is the equation (and system)? What system parameter(s) characterize the system?

Problem 3 You're escaping the East India Trading Company in your trusty vessel "The Black Pearl." The Pearl's sails generate thrust in the following relationship:

$$F_{Sail} = C_S(V_W - V_p)$$

where V_P is the velocity of the Pearl, V_W is the velocity of the wind, and C_S is a constant. The drag on the Pearl's hull is linearly proportional to her velocity:

$$F_{Drag} = C_D V_P$$

where C_D and the Pearl's mass, m, are constant.

Use an appropriately labeled free body diagram to derive the differential equation governing the Pearl's velocity. Determine an algebraic expression for the Pearl's terminal, i.e., steady state, velocity. Identify the time constant, τ, for the ship's velocity response.

Problem 4 Consider the mass-spring-damper system subjected to a ramp input platform displacement, $y(t) = 5t$ as shown:

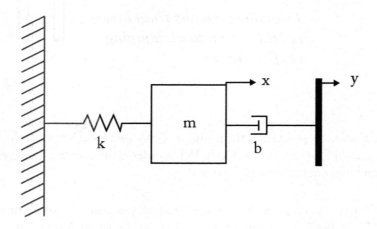

(a) Draw an appropriately labeled free body diagram and derive the governing differential equation for the displacement of the mass.

(b) What order are the equation and the system?

(c) What is/are the relevant system parameter(s)?

Problem 5 Consider the downhill skier pictured here:

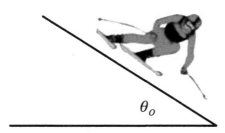

The total drag on the skier, F_D, is a combination of man-made-snow surface resistance and aerodynamic drag resulting in the following relationship for the drag force: $\vec{F}_D = C_D \vec{V}$ where C_D is the coefficient of drag, $\vec{V}$ is the velocity of the skier down the inclined slope, and $C_D = $ constant. Draw an appropriately labeled free body diagram and derive the equation governing the skier's velocity. Determine the relevant system parameter(s) for the model.

Problem 6 A pressure compensating hydraulic spool valve consists of a bar-bell-like mass in a cylindrical sleeve (shown below). The valve is moved horizontally by a solenoid that applies a step input force to the mass. A spring at the far end provides an opposing force. Hydraulic fluid in a tight clearance of width, h, provides a viscous friction force resisting the motion and given by the relation:

$$F_v = \frac{Cv}{h}$$

where C is a constant.

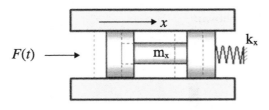 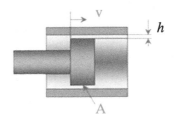

Show that a balance of forces in the horizontal direction gives:

$$m\frac{d^2x}{dt^2} = F(t) - kx - \frac{Cv}{h}.$$

This equation physically represents a statement of what balance principle? Write algebraic expressions for the system natural frequency and damping ratio in terms of the provided quantities.

Problem 7 Consider the angular position of a 100 kg winter Olympic snowboarder on a circular pipe of radius, R. The total drag on the snowboarder, F_D, is a combination of man-made-snow surface resistance and aerodynamic drag resulting in the following relationship for the drag force: $\vec{F_D} = C_D \, \vec{V}$ where C_D is the coefficient of drag and $\vec{V}$ is the tangential velocity of the snowboarder and $C_D =$ constant. Use $I = mR^2$.

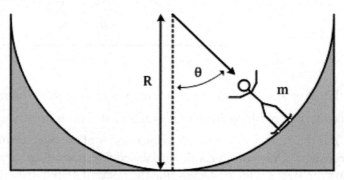

Using an appropriately labeled free body diagram and applying a balance of torques, show that the differential equation governing the angular position of our snowboarder with respect to time is given by

$$mR^2\ddot{\theta} + C_D R^2 \dot{\theta} + mgR \sin\theta = 0.$$

If the skier could enter the pipe at an angle of 30 degrees and remain at angles equal to or lower than this, linearize the equation to obtain a linear, ordinary differential equation governing the skier's angular position.

Problem 8 Consider the mass-less-platform-spring-damper system subjected to a ramp input platform displacement, $y(t) = 5t$ as shown:

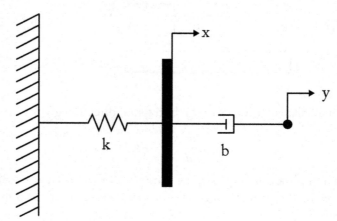

(a) Draw an appropriately labeled free body diagram and derive the governing differential equation for the displacement of the mass.

(b) What order are the equation and the system?

(c) What is/are the relevant system parameter(s)?

CHAPTER 5

A Common Notion

Euclid's first common notion is this: things that are equal to the same thing are equal to each other. That's a rule of mathematical reasoning. It's true because it works. Has done and always will do. Euclid says this truth is self-evident. You see … there it is, even in that 2000 year old book of mechanical law. It is a self-evident truth that things which are equal to the same thing are equal to each other.

Abraham Lincoln quoting Euclid's
Book of Common Notions

I understand what an equation means if I have a way to figure out the characteristics of its solution without solving it.

Richard P. Feynman
quoting Paul Dirac

Mr. Lincoln read Euclid wisely: two things equal to the same thing are equal to each other. This basic premise lies at the notion of a common solution for linear ordinary differential equations. What we will learn here is that all solutions for first order systems look like "the same thing." This will also hold for all 2nd order systems. In Chapters 2 and 3, we motivated the independent physical principles of inertia, stiffness, and friction (or alternatively inductance, capacitance, and resistance) by linking them with a cartoon-like characterization in an attempt to illustrate the analogous roles these play in mechanical and electrical systems. We further made this characterization to create a mnemonic device by which the abstract mathematics used to model such systems may be more approachable and less daunting. In fact, because the mathematics is essentially "always the same thing," the analogy serves to teach us that there's less to learn than we might otherwise have thought.

We further associated principles of inertia, stiffness, and friction with their physical roles as agents of storage of kinetic energy, potential energy, and the dissipation of energy, respectively. We then followed a universal principle of balancing or conserving a basic quantity entering and leaving a volume element of the system. When we introduce mathematical relations for Captains Potential and Kinetic Energy and the Evil Dr. Friction using effort and flow variables, it is a relatively painless procedure to write a governing ordinary differential equation for a system. So far, the common axioms of systems in different disciplines are:

(a) Each system contains elements represented by characters that:

(a) Store kinetic energy, e.g., inertia or inductance

(b) Store potential energy, e.g., stiffness or capacitance

(c) Dissipate energy, e.g., friction or electrical resistance.

(b) The governing differential equation results from expressing conservation or balance of elemental quantities, e.g., momentum in a mechanical system, or change through a representative electrical circuit node, and

(c) Very specific and critically important quantities called system parameters arise out of various ratios, products, and sums of the system elements, e.g.,

(a) The time constant, τ, for linear, 1st order, ordinary differential equations

(b) The natural frequency, ω_N, and the damping ratio, ζ, for linear, 2nd order, ordinary differential equations.

What makes these quantities so crucial is that they characterize everything interesting about the mathematical solutions. In the following sections, we will discuss and dissect these solutions for linear, first and second order differential equations in terms of the system parameters. Remember, the specific mathematical form of the system parameters, the time constant, natural frequency, and damping ratio, arise from the individual discipline-specific actors playing out a common movie script.

5.1   TIME DOMAIN SOLUTIONS OF 1st ORDER SYSTEMS

Consider the movie scripts we discussed in Chapter 2 that correspond to 1st order systems. First order systems result when the script involves a single *type* of storage element or character (either potential or kinetic energy storage) along with dissipative elements. Note there may be multiple agents of storage, but they must store only one type of energy. So far, we've been introduced to:

(a) A single electrical capacitor with a resistor, e.g., a series RC circuit with battery

(b) A single electrical inductor with a resistor, e.g., a series RL circuit with battery

(c) A single mechanical spring with a dashpot or friction element arranged in parallel, e.g., the idealized, mass-less spring-dashpot system

(d) A single mechanical inertia with a friction element, e.g., the parachutist in free-fall.

In all these cases, the governing differential equation has the form shown in Section 3.4.3:

$$\tau \frac{d\psi(t)}{dt} + \psi(t) = \Psi(t)$$
$$\psi(t = 0) = \psi_O$$

Table 5.1: Relation of the system time constant to the system element parameters

System	Time Constant	Electrical Analogy
Series RC Circuit	R*C	
Series RL Circuit	L/R	
Massless Spring Dashpot	b/k	Product $R_{MECH} * C_{MECH}$
Freefall Parachutist	m/b	Ratio L_{MECH} / R_{MECH}

where ψ is either effort or flow that is stored in the system and the time constant, τ, depends on the individual inertial, stiffness, or friction quantities.

Therefore, one needs only identify the storage and dissipative elements and their structural arrangement to conclude the relevant time constant. Recall, this is illustrated in Figure 2.4. The system response for $\psi(t)$ is then driven by the system's initial condition and the forcing function or signal input, $\Psi(t)$.

For linear systems, solutions for $\psi(t)$ may be obtained by either use of Laplace transforms in the complex plane or a superposition of homogeneous and complimentary solutions in the time domain. Laplace transform solutions are available in a number of good texts on systems dynamics [10, 11, 19]. For the purposes of physical interpretation, we choose here to restrict ourselves to solutions strictly in the time domain. By doing this, we hope to replace the mathematical jargon with the physical meaning underlying the math.

From courses in elementary differential equations, we recall that any linear, ordinary, first order differential equation in a single independent variable exhibits a solution that can be posed as the sum of the response, $\psi_h(t)$, to the corresponding homogeneous differential equation

$$\tau \frac{d\psi_h(t)}{dt} + \psi_h(t) = 0$$

and the particular response, $\psi_p(t)$, to the differential equation driven by the external signal input or forcing function, $\Psi(t)$:

$$\tau \frac{d\psi_p(t)}{dt} + \psi_p(t) = \Psi(t)$$

where the total solution, via linear superposition, is given by:

$$\psi(t) = \psi_h(t) + \psi_p(t).$$

You probably were shown this in your earlier courses in differential equations. The *homogeneous* solution is often referred to as the *natural* or *free* response as this portion of the solution solves the equation where only the system parameters appear and there is no forcing function or agent of change from the outside world. Father Force is AWOL in this part of the response. It's all about the system on the left side of the equation. Recall this from the illustration given by the character equation in Figure 2.4 for the RC circuit. This part of the solution prescribes how the system will react when free of external forces or inputs, i.e., how a system responds essentially to initial conditions. Therefore, the natural response will be a function of the system parameters ONLY, i.e., in the case of a first order system, the time constant, τ.

The portion of the solution that responds directly to the excitation from the outside world is the so-called particular solution. An *agent external to the system* is forcing the system to respond to it. We can understand this distinction even more clearly once we have solved both differential equations.

5.1.1 TRANSIENT RESPONSE

> Mathematicians postulate forms for solutions to differential equations ... well, let's face it, they guess.

P.E. Wellstead
Introduction to Physical System Modeling

While there is, of course, more to it than that, we, as engineers, rather than mathematicians, are happy to take the nod on the form of the solution. Many real physical systems exhibit exponential behavior. They can be modeled as first order ordinary differential equations because an exponential solution works to "solve" it.

$$\psi_h(t) = Ae^{\lambda t}$$

where the unknown quantity, λ, results from satisfying the homogeneous form of the governing differential equation:

$$\tau A\lambda e^{\lambda t} + Ae^{\lambda t} = 0.$$

Dividing through by $Ae^{\lambda t}$ renders the characteristic equation:

$$\tau\lambda + 1 = 0 \Rightarrow \lambda = -1/\tau.$$

So solutions like $\psi_h(t) = Ae^{\lambda t}$ work when $\lambda = -1/\tau$. So we have

$$\psi_h(t) = Ae^{-1/\tau}.$$

The value of the constant, A, is determined by applying system's initial conditions after the complete or total solution is found. The natural response is an exponential decay over the dimensionless time, t/τ, and represents the part of the solution that responds to the system's initial conditions.

Of Special Note

The free response is a response to initial conditions in the absence of any external forcing from the outside world. We can associate this response with the transient response of the system. As it is purely exponential in nature, it "dies out" in a finite amount of time we call the settling time.

5.1.2   FORCED RESPONSE

The mathematical particular solution, $\psi_p(t)$, responds directly to the forcing function imposed by the outside world. The proper form for this response is a function that is, in some sense, the most general form of the function driving the system. Some familiar forms of input excitations are shown in Table 5.2.

Table 5.2: General forms of particular solutions corresponding to a variety of polynomial input excitations

Input Excitation	General Form of $\psi_p(t)$
Step	Constant: K
Ramp or Step-Ramp	Linear: Ct + K
Polynomial	Similar Order Polynomial $\psi_p(t) = At^N + Bt^{N-1} + \ldots + Ct + K$
Harmonic $\psi_O(t) = A_{IN} \cos(\omega t + \alpha)$	Harmonic $\psi_p(t) = A_{OUT} \cos(\omega t + \alpha + \varphi)$
Arbitrary Function	Truncated Polynomial Taylor Series

We can more clearly show the physical interpretation of the forced response by performing a full solution for a few simple examples.

Step Input

Consider the example of the mass-less plate discussed in Section 4.3 wherein a constant force is instantaneously applied to the plate and maintained:

$$F_0(t) = \begin{cases} 0 & t < 0 \\ P & t \geq 0 \end{cases}$$

for which the appropriate forced response is a constant:

$$\psi_p(t) = K = \text{constant}.$$

This function must now satisfy the inhomogeneous or forced version of the differential equation

$$\tau\frac{d}{dt}(K) + K = P/k$$

$$K = P/k.$$

So the particular solution is that amount of deformation the spring would experience under a purely static load P, i.e., $P/k = \delta_{STATIC}$. The forced system response is then simply the static deflection of the spring alone. To understand this in more detail, let's compose the total solution for the position of the plate

$$x(t) = x_h(t) + x_p(t) = Ae^{-t/\tau}P/k$$

$$x(0) = x_0 = A + P/k$$

$$\Rightarrow A = x_0 - P/k$$

or

$$x(t) = (x_0 - P/k)e^{-t/\tau} + P/k.$$

Notice that since the transient, by definition, decays away at long times compared with the system time constant, τ, the particular solution must represent that part of the solution that remains at long times or the steady state. This solution is shown graphically in Figure 5.1.

We learn several interesting characteristics from this response that, it turns out, are characteristic of all first order responses. Since for our case:

$$x(0) = x_0 = 2\,\text{m}$$

$$x(t \geq 4\tau) \approx P/k = x_{SS} = 12\,\text{m}$$

$$x(t) = (x_0 - P/k)e^{-t/\tau} + P/k.$$

The response to the step input force proceeds exponentially from the initial value of 2 meters to the final value of $x_{SS} = P/k = \delta_{STATIC}$ in approximately four time constants. As engineers, we choose a somewhat arbitrary datum for the time at which the exponential decay is sufficiently complete. Here, we adopt as a reference point the time by which 98% of the change from the initial value to the steady-state value takes place. This is four time constants because 98% of the exponential decay has occurred within this time frame:

$$e^{-t/\tau} = e^{-4\tau/\tau} = e^{-4} = 0.018 \approx 0.02.$$

We often refer to this regime as the *transient* because the plate position is changing throughout this time range. Thereafter the response is in the *steady-state* at a value equaling that given by the static deflection of the spring alone because the plate is effectively no longer moving and the internal force in the damper has decayed to some negligibly small value.

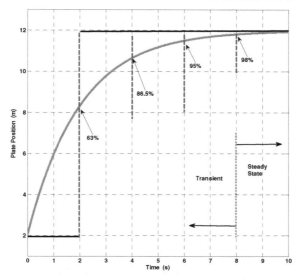

Figure 5.1: The response of a mass-less plate with spring and damper to a step input force of 60 N ($x_0 = 2$ m; $k = 5$ N/m; $b = 10$ Ns/m). The time constant is given by $b/k = 2$ seconds. The response is characterized by an exponential approach from an initial value to a final value of $\delta_{STATIC} = P/k$.

Of Special Note

> *The forced response is a response specifically to the external forcing from the outside world. This response is present long after the transient or free response has decayed away. For this reason, the forced response is often referred to as the steady-state response.*

When one examines these regimes along with the mathematics of the homogeneous and particular solutions, one can list several observations that are universally true for all first order systems.

Of Special Note

Observations regarding solutions to all 1st order differential equations

(a) The homogenous solution *responds to the initial conditions* and represents the mathematical structure of the physical *transient* from initial to steady-state values.

(b) The particular solution *responds specifically to the forcing function* imposed upon the system by some external agent. It is the only portion of the solution that survives after the exponential decay of the transient. As such, $\psi_p(t)$ represents the response of the system in *steady state*.[a]

(c) In the parlance of a movie script, from beginning (initial) to end (steady state) values, the transient part of the movie lasts roughly 4 time constants. Admittedly, this number is somewhat arbitrary and can be adjusted to please the precision with which one needs to attain steady state. What doesn't change is that the steady state is effectively attained in quanta of time constants

(d) Lastly, the entire response can be cast in dimensionless form. This is always an appealing feature in predictive models because it points toward physically motivated model parameters.

---

[a]In this language, the steady state is generally a function of time when the input signal is time-dependent.

To see this last point, one can reformulate the solution to take the form of a dimensionless response variable, $\hat{\psi}(t)$ where

$$\hat{\psi}(t) = \frac{\psi(t) - \psi_{SS}}{\psi_0 - \psi_{SS}} = e^{-\hat{t}} \quad \text{where} \quad \hat{t} = \frac{t}{\tau}$$

which is plotted in Figure 5.2. Often, students will only first see this dimensionless form of solutions to first order differential equations in their undergraduate heat transfer courses. As you may not have yet had such a course, what is important to point out is that the term $\psi(t) - \psi_{SS}$ is the driving agent that causes the variable $\psi(t)$ to change over time. When the variable eventually reaches its steady-state value, this driving agent vanishes and the transient is complete. So the main "take away" concept here is that the driver for dynamic response is the measure by which the current value of the variable is different from its eventual steady-state value. It is precisely this difference in values that actually exponentially decays away in time. Because all systems, regardless of their initial conditions or forcing function, can be cast in this form, we can refer to this dimensionless form as a master curve for first order systems. A master curve is a function onto which all solutions fall when appropriately normalized or non-dimensionalized. Master curves are appealing in predictive mathematical modeling because of the physical interpretation given to the normalizing quantities. Here, these are the difference between the value of the dependent system variable and its eventual steady-state value, i.e., the driving force, and the system time constant.

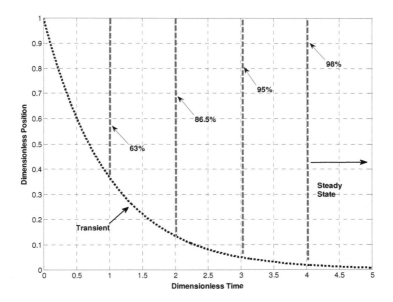

Figure 5.2: The response of a mass-less plate with spring and damper to a step input force in dimensionless form.

The difference between a system response variable, $\psi(t)$, and its value in steady state is the driver causing the dynamic response. As for most meaningful dimensionless parameters in models, $\hat{\psi}(t)$ represents a ratio between two physical quantities: the ratio of the current driving agent to the initial driving agent. Therefore, this particular ratio of differences always decays exponentially in first order systems over time regimes measured in quanta of system time constants.

Ramp Input
We can maintain that the generalization holds when the system is exposed to a time-dependent forcing function. Consider the ramp input signal:

$$F_0(t) = \begin{cases} 0 & t < 0 \\ 10t & t \geq 0 \end{cases}$$

for which the appropriate normalized signal input in our example is $F(t)/k = 10t/k$. The most general form of a linear function is then presumed for the particular solution:

$$\psi_p(t) = Ct + K.$$

Substituting this into the differential equation

$$\tau C + K + Ct = (10/k)t$$

and upon setting like terms equal to one another:

$$\tau C + K + Ct = (10/k)t$$
$$C = 10/k$$
$$0 = C\tau + K$$
$$K = -C\tau = \frac{-10}{k}\tau$$

or

$$\psi_p(t) = -\frac{10}{k}\tau + \frac{10}{k}t = \frac{10}{k}(t - \tau).$$

It is important to note that while the forcing function is a straight line with zero intercept the eventual steady-state solution has an intercept. This implies there is an offset in time between the forcing function and the steady response. This steady solution given by $\psi_p(t) = \frac{10}{k}(t - \tau)$ is the straight dotted line in Figure 5.3.

Compiling the total solution and applying the initial conditions:

$$x(t) = x_h(t) + x_p(t) = Ae^{-t/\tau} + 10(t - \tau)/k$$
$$x(0) = x_0 = A - 10\tau/k$$
$$x(t) = (x_0 + 10\tau/k)e^{-t/\tau} + 10(t - \tau)/k$$

which is plotted in Figure 5.3 for several distinct initial displacements along with the asymptotic steady-state line.

5.1.3   DIMENSIONLESS SOLUTIONS FOR 1st ORDER SYSTEMS

We note that even when the steady state is time-dependent, the entire response can still, for *every* first order system, be cast in dimensionless form:

$$\hat{\psi}(t) = \frac{\psi(t) - \psi_{SS}(t)}{\psi_0 - \psi_{SS}(0)} = e^{-\hat{t}} \quad \text{where} \quad \hat{t} = \frac{t}{\tau}.$$

This dimensionless solution is plotted in Figure 5.4. Note the form is identical with that in Figure 5.2.

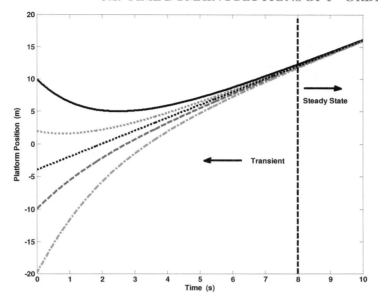

Figure 5.3: The response of a mass-less plate with spring and damper to a step input force in dimensional form. The response is characterized by an exponential approach or transient from an initial value to a final value that, like the forcing function, increases linearly in time.

5.1.4 UNIVERSAL TRUTHS FOR 1st ORDER SYSTEM RESPONSE IN THE TIME DOMAIN

We can now add several observations to our list of universal truths that always characterize how 1st order systems respond to their environment. We note that 1st order systems always approach a steady-state response monotonically from their initial condition, and the response never overshoots this steady response. The steady response behaves like "a fence" that bounds the total response. This total response approaches the steady solution "from one side" where the initial conditions reside, as observed in Figure 5.3 for the ramp input example. We also note that even when the steady-state solution is time-dependent, the appropriate non-dimensionalization delivers a master curve that is identical for all initial conditions, or starting points, and steady-state solutions or ending points as shown in Figure 5.4.

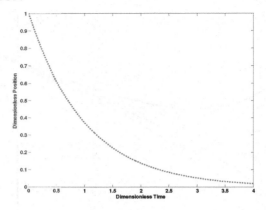

Figure 5.4: The response of a mass-less plate with spring and damper to a ramp input force in dimensionless form.

Of Special Note
   Universal Truths for 1ˢᵗ Order Systems

(a) They are comprised of system elements (or characters) that store only a single form of energy, *either* potential *or* kinetic energy (but not both).

(b) Their behavior is characterized by a single system parameter called the system time constant, τ, where

$$\tau = f_1\,(R, C) = f_2\,(b, k) \quad \text{or} \quad \tau = g_1\,(L, R) = g_2\,(m, b)\,.$$

(c) The system is identified with one characteristic time given by the time constant.

(d) The system transient decays away in a multiple number of system time constants.

(e) The system response approaches steady state monotonically from one side.

(f) The system is *never* capable of overshooting the eventual steady-state response.

(g) The system response can be universally placed in a dimensionless form normalized by the driving agent $\psi - \psi_{SS}$, and the characteristic time constant, τ.

5.2   TIME DOMAIN SOLUTIONS OF 2nd ORDER SYSTEMS

Again, as discussed in detail in Chapter 2, 2nd order systems result when the script involves both distinct *types* of storage character, i.e., *both* potential *and* kinetic energy storage. Note there may be multiple agents of storage, but they must be capable of storing both types of energy. So far, we've been introduced to:

(a) An electrical capacitor and inductor with a resistor, e.g., series or parallel RLC circuits with an external passive power supply, i.e., battery.

(b) An electrical capacitor and inductor with no systemic damping, e.g., a series/parallel LC circuit with an external passive power supply, i.e., battery.

(c) A mass with mechanical spring and dashpot connected in series or parallel, e.g., the idealized, mass-spring-dashpot system.

(d) An idealized, undamped mass-spring harmonic oscillator.

In these cases, the normalized governing differential equation has the form:

$$\frac{1}{\omega_N^2}\frac{d^2\psi(t)}{dt^2} + \frac{2\zeta}{\omega_N}\frac{d\psi(t)}{dt} + \psi(t) = \Psi(t)$$

where ψ reresents the dependent effort or flow variable in the system and the system is characterized by a pair of parameters: the damping ratio and natural frequency, $\{\zeta, \omega_N\}$, where each depends on the individual inertial, stiffness, or friction quantities. Examining the systems in Sections 3.5 and 4.5, we arrived at the results summarized in Table 5.3.

When viewed from the perspective of the effort-flow analogy with electrical systems, i.e., considering the system damping ratio and natural frequency, a *parallel* mass-spring-damper system should behave similarly to the *series* RLC circuit (see Table 5.3). Therefore, one need only identify the storage and dissipative elements and how they are structured in the system to know that the relevant natural frequency and damping ratio are particular products and/or ratios of the respective system element parameters. Recall, this is illustrated in Tables 3.3 and 3.4 for 1st order electrical systems. The system response for $\psi(t)$ is driven by one or both of the system's initial conditions and the forcing function or signal input, $\Psi(t)$.

The corresponding transient and steady-state parts of the solution satisfy:

$$\frac{1}{\omega_N^2}\frac{d^2\psi_h(t)}{dt^2} + \frac{2\zeta}{\omega_N}\frac{d\psi_h(t)}{dt} + \psi_h(t) = 0$$

and

$$\frac{1}{\omega_N^2}\frac{d^2\psi_p(t)}{dt^2} + \frac{2\zeta}{\omega_N}\frac{d\psi_p(t)}{dt} + \psi_p(t) = \Psi(t)$$

Table 5.3: Analogous representations for system natural frequency and damping ratio

System	Natural Frequency (rad/s)	Electrical Analogy	Damping Ratio	Electrical Analogy
Series RLC Circuit	$\omega_N = 1\big/\sqrt{LC}$		$\zeta = \dfrac{R}{2}\sqrt{\dfrac{C}{L}}$	
Parallel RLC Circuit	$\omega_N = 1\big/\sqrt{LC}$		$\zeta = \dfrac{1}{2R}\sqrt{\dfrac{L}{C}}$	
Parallel Mass-Spring-Damper	$\omega_N = \sqrt{\dfrac{k}{m}}$	$\omega_N = \dfrac{1}{\sqrt{L_{MECH} * C_{MECH}}}$	$\zeta = \dfrac{b}{2\sqrt{km}}$	$\zeta = \dfrac{R_{MECH}}{2}\sqrt{\dfrac{C_{MECH}}{L_{MECH}}}$
Parallel Mass-Spring	$\omega_N = \sqrt{\dfrac{k}{m}}$	$\omega_N = \dfrac{1}{\sqrt{L_{MECH} * C_{MECH}}}$	$\zeta = 0$	$\zeta = 0$

respectively, where the total solution, via linear superposition, is given by:

$$\psi(t) = \psi_h(t) + \psi_p(t).$$

The transient response will be a function of the system parameters only, i.e., the system natural frequency, ω_N, and system damping ratio, ζ, and is that portion of the solution that responds directly to the initial conditions. The portion of the solution that responds directly to the excitation from the outside world is the particular solution, $\psi_p(t)$.

5.2.1   FREE RESPONSE

Similar to first order equations, exponential functions also satisfy the second order equation:

$$\psi_h(t) = Ae^{\lambda t}$$

where the unknown exponents result from satisfying the ODE:

$$\frac{1}{\omega_N^2} A\lambda^2 e^{\lambda t} + \frac{2\zeta}{\omega_N} A\lambda e^{\lambda t} + Ae^{\lambda t} = 0.$$

Dividing through by $Ae^{\lambda t}$ renders the characteristic equation for the ODE:

$$\frac{1}{\omega_N^2}\lambda^2 + \frac{2\zeta}{\omega_N}\lambda + 1 = 0 \quad \Rightarrow \quad \text{two solutions, } \lambda_{1,2}.$$

Because this equation is quadratic, it exhibits two roots, $\lambda_{1,2}$. Depending on the sign of the discriminant, pairs of roots to this equation correspond to three distinctly different physical regimes of behavior as in Table 5.4.

Table 5.4: The three physical scenarios for transient solutions of second order systems

Scenario	ζ	Nature of roots	Roots	Physical Regime
1	$\zeta < 1$	Pair of complex conjugate roots	$\lambda_{1,2} = -\zeta\omega_N \pm j\omega_N\sqrt{1-\zeta^2}$	Under Damped
2	$\zeta = 1$	Pair of two real, equal roots	$\lambda_1 = \lambda_2 = -\zeta\omega_N$	Critically Damped
3	$\zeta > 1$	Pair of two real, distinct roots	$\lambda_{1,2} = -\zeta\omega_N \pm \omega_N\sqrt{\zeta^2-1}$	Over Damped

Underdamped Systems

For the first scenario, the system is under-damped. Mathematically, these result when the damping ratio, ζ, is less than one. Physically, this happens when elasticity and inertia dominate friction in a system (see Figure 5.5). Recall, $\zeta = b/2\sqrt{km}$. If the stiffness and inertia, $\sqrt{km}$ dominate relative to dissipation, b, then $\zeta < 1$. Energy storage, in some sense, is strong enough to overcome energy dissipation allowing for a transfer back and forth between potential and kinetic forms of energy in the system.

Figure 5.5: Strength of energy characters in an under-damped 2nd order system.

This is rendered mathematically by the solutions of the characteristic equation. The transient solution is given by:

$$\psi_h(t) = C_1 e^{\lambda_1 t} + C_2 e^{\lambda_2 t}$$

where C_1 and C_2 are constants to be determined later by imposing the initial conditions. When the roots are complex, they contain a negative real portion corresponding to the exponential decay caused by the energy dissipating character, and a purely imaginary portion that corresponds to a harmonic oscillation that occurs "inside the decaying envelope" of the exponential part of the solution. These are the energy storage characters transferring energy back and forth between potential and kinetic forms. Algebraic manipulation results in a solution of the form

$$\psi_h(t) = e^{-\zeta\omega_N t}\left[A\cos(\omega_d t) + B\sin(\omega_d t)\right]$$

where $\omega_d = \omega_N\sqrt{1-\zeta^2}$ is the damped natural frequency, and A and B are constants determined by applying the initial conditions. The system will oscillate in the transient with this characteristic frequency in the presence of energy dissipation. And the natural response decays away exponentially in a time frame prescribed by the system damping ratio and natural frequency. There is no time constant, per se, for a second order system. The constants, A and B, are determined by applying the system's initial conditions. The under-damped transient response is an exponentially decaying harmonic that decays over the dimensionless time, $\hat{t} = t/(1/\zeta\omega_N)$.

Of Special Note

It is interesting to note that while resistance is fairly straightforward to quantify in electrical systems, damping coefficients in mechanical systems have a somewhat higher degree of uncertainty associated with them. You will not find a value for the damping coefficient stamped on the container for a damping element. Friction always has an inherent uncertainty about its actual mathematical representation.

Because of this, and because it is the damping ratio, ζ, and not the damping coefficient, b, that matters in our solutions, we point out that there is a straightforward way to determine the damping ratio directly from experimental data. For this purpose, imagine that you perturb an under-damped second order system from rest with an initial displacement and let the system's free response decay away from the rest. It is easy to show that the ratio of successive peaks is given by:

$$\frac{x_{N+1}}{x_N} = e^{-\zeta\omega_N T_d}$$

$$\delta \equiv \ln\left(\frac{x_{N+1}}{x_N}\right) = \zeta\omega_N T_d = \frac{2\pi\zeta}{\sqrt{1-\zeta^2}}.$$

Thereby, the log decrement, δ, a quantity readily measured from experiment, is a function solely of the system's damping ratio. Inverting this relation

$$\zeta = \frac{\delta}{\sqrt{4\pi^2 + \delta^2}}.$$

So the damping ratio is easily determined by measuring the log decrement, δ, or logarithm of ratios of successive peaks. The damped period of the free decaying oscillation is also easy to measure. With the period known, it is straightforward to compute the system's natural frequency:

$$\omega_N = 2\pi / T_d \sqrt{1 - \zeta^2}.$$

So the system parameters can be computed directly from simple experimental measurements.

Critically Damped Systems

The second scenario is basically a fence between the 1st and 3rd scenarios. The system is called critically damped. Physically, this corresponds to a system where the energy storage and dissipation "have equal strength," if you will, and $b = 2\sqrt{km}$ (see Figure 5.6). The ability of the system's elasticity and inertia to store potential and kinetic energy, respectively, is "equal," in some sense, to the ability of the system to dissipate energy. Energy storage, then, just rivals energy dissipation. In this limit, there is just enough friction or dissipation to prevent anything other than a single transfer of energy between the kinetic and potential forms. As such, there is sufficient enough energy dissipation to just prevent oscillatory response from occurring.

Figure 5.6: Strength of energy characters in a critically damped 2nd order system.

The solutions contain two negative, equal real parts, $\lambda_1 = \lambda_2 = \lambda = -\zeta\omega_N$, corresponding to the exponential decay caused by the energy dissipating characters. The transient solution is given by:

$$\psi_h(t) = C_1 e^{\lambda t} + C_2 t e^{\lambda t} = C_1 e^{-\zeta\omega_N t} + C_2 t e^{-\zeta\omega_N t}.$$

The constants, C_1 and C_2, are determined by the system's initial conditions. The critically damped transient response is a pure exponential decay over the dimensionless time, $\hat{t} = t/(1/\zeta\omega_N)$. As we will soon observe, this decay is the fastest decay that does not allow for oscillatory behavior in the transient. This makes the critically damped response case an important limit solution for engineering design as there are a number of physical situations in which one desires as fast a decay as possible without oscillation from a given prescribed set of initial conditions, e.g., response of a mass-spring-damper automobile strut to an imposed initial compression.

Overdamped Systems

Physically, the final scenario corresponds to a system where the energy dissipation dominates the response at the expense of the ability of the system's elasticity and inertia to store potential and kinetic energy respectively. Mathematically, $b > 2\sqrt{km} \Rightarrow \zeta > 1$. In this limit, there is more energy dissipation than is necessary to prevent oscillatory response from occurring (see Figure 5.7).

Figure 5.7: Strength of energy characters in an overdamped 2nd order system.

This is rendered mathematically by the solutions of the characteristic equation: two negative, and distinct real parts, $\lambda_{1,2} = -\zeta\omega_N \pm \omega_N\sqrt{\zeta^2 - 1}$ corresponding to two distinct rates of exponential decay caused by the energy dissipating characters. The transient solution is given by:

$$\psi_h(t) = C_1 e^{\lambda_1 t} + C_2 e^{\lambda_2 t} = C_1 e^{\left(-\zeta\omega_N + \omega_N\sqrt{\zeta^2-1}\right)t} + C_2 e^{\left(-\zeta\omega_N - \omega_N\sqrt{\zeta^2-1}\right)t}.$$

The constants, C_1 and C_2, are determined by the system's initial conditions. The critically damped transient response is a pair of pure exponential decays over two distinct dimensionless times:

$$\hat{t}_1 = t / \left(1/\left(-\zeta\omega_N + \omega_N\sqrt{\zeta^2 - 1}\right)\right) \quad \text{and} \quad \hat{t}_2 = t / \left(1/\left(-\zeta\omega_N - \omega_N\sqrt{\zeta^2 - 1}\right)\right).$$

The over damped response is identified by two physical time scales for decay:

(a) one decay time that is larger than that in the critically damped case

(b) a distinct second decay time that is smaller than that in the critically damped case.

Thus, superposing both solutions results in an overall decay time longer than that observed in the critically damped case. The more damping or friction added to a system beyond this limit, the slower the decay to steady state.

5.2.2   FORCED RESPONSE

The handling of the mathematical particular solution, $\psi_p(t)$, is no different than that for 1st order systems. The solution to the inhomogeneous differential equation responds directly to the forcing function. The form for this response is the most general form of the function driving the system as outlined in Table 5.2. Again, the physical interpretation of the forced response is best shown by performing several simple examples.

Step Input to an Underdamped System

Consider the example of the parallel mass-spring-damper discussed in Section 4.4 wherein a constant force is instantaneously applied. The classic step input signal is simply a constant input suddenly applied:

$$F_0(t) = \begin{cases} 0 & t < 0 \\ P & t \geq 0 \end{cases}$$

for which the appropriate forced response is

$$\psi_p(t) = K = \text{ constant.}$$

This function must now satisfy the inhomogeneous or forced version of the ODE

$$\frac{1}{\omega_N^2}\ddot{K} + \frac{2\zeta}{\omega_N}\dot{K} + K = \frac{P}{k}$$

$$K = P/k = \delta_{STATIC}.$$

The forced system response is then simply the static deflection of the spring alone. To understand this in more detail, let's compose the total solution for the position of the plate

$$x(t) = x_h(t) + x_p(t) = e^{-\zeta\omega_N t}\left[A\cos(\omega_d t) + B\sin(\omega_d t)\right] + P/k.$$

Applying the initial conditions:

$$x(0) = x_0 \Rightarrow A = x_0 - P/k$$

$$\dot{x}(0) = v_0 \Rightarrow B = \frac{v_0 + \zeta\omega_N(x_0 - P/k)}{\omega_N\sqrt{1-\zeta^2}}.$$

Finally, upon substitution

$$x(t) = e^{-\zeta\omega_N t}\left[(x_0 - P/k)\cos(\omega_d t) + \frac{v_0 + \zeta\omega_N(x_0 - P/k)}{\omega_N\sqrt{1-\zeta^2}}\sin(\omega_d t)\right] + P/k.$$

This solution is shown graphically in Figure 5.8 for several sets of initial displacements.

This response exhibits several features characteristic of all under-damped 2nd order responses: an exponentially decaying, oscillatory, harmonic response that overshoots the eventual

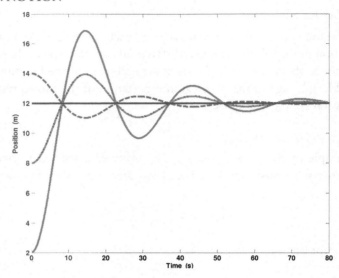

Figure 5.8: The response of a lumped mass with parallel spring and damper to a step input force of 60 N ($x_0^{1,2,3}$ = 2 m, 8 m, 14 m; v_0 = 0 m/s; k = 5 N/m; b = 10 Ns/m). The response is characterized by a decaying oscillation from an initial value that overshoots its steady-state value. It oscillates about and ultimately decays to the steady-state value of δ_{STATIC} = F_0/k.

steady-state solution at x_{SS} = F_0/k = δ_{STATIC}. This eventual steady state is then reached in approximately four characteristic decay times parameterized by ω_N, and ζ. This is the *transient* regime where the response is characterized by two characteristic times: the decay time of the envelope bounding the oscillations and the period of the oscillations as outlined in Table 5.5.

Table 5.5: Characteristic transient time scales for under-damped second order systems

Solution Feature	*Characteristic Time*
Exponential decay	$1/\zeta\omega_N$
Period of damped harmonic response	$2\pi/\omega_d = 2\pi/\omega_N\sqrt{1-\zeta^2}$

Following the exponential decay, the response is in the *steady state* at a value equaling that given by the static deflection of the spring alone because the inertial mass is effectively no longer moving and the internal force in the damper has decayed to some negligibly small value. Again, precisely as for first order systems, we observe characteristics common to the solutions of all 2nd order systems.

Of Special Note

Observations regarding solutions to all 2nd order differential equations

(a) The homogenous solution **responds to the initial conditions** and represents the mathematical structure of the physical transient from initial to steady-state values.

(b) The particular solution **responds specifically to the signal input or forcing function** imposed upon the system by some external agent, i.e., the outside world. It is the only portion of the solution that survives after the exponential decay of the transient. As such, $\psi_p(t)$ represents the response of the system in steady state.

(c) In the parlance of a movie script, from beginning (initial) to end (steady-state) values, the movie lasts effectively 4 characteristic decay times as prescribed in Table 5.5. So the steady state is effectively attained in quanta of exponential decay times.

(d) Lastly, the entire response can, for **every** second order system, be cast in dimensionless form.

To see this last point, one can reformulate the solution to take the form of a dimensionless response variable, $\hat{\psi}(t)$ where

$$\hat{\psi}(t) = \frac{\psi(t) - \psi_{SS}}{\psi_0 - \psi_{SS}} = Ge^{-\hat{t}_1} \cos\left(2\pi\hat{t}_2 - \tan^{-1}\left\{\frac{\zeta(\eta+1)}{\sqrt{1-\zeta^2}}\right\}\right)$$

$$G = \sqrt{\frac{1 + 2\eta\zeta^2 + \eta^2\zeta^2}{1 - \zeta^2}}$$

$$\eta = \frac{v_0}{\zeta\omega_N(x_0 - x_{SS})}$$

$$\hat{t}_1 = t/(1/\zeta\omega_N)$$

$$\hat{t}_2 = t/(2\pi/\omega_d) = t/T_d$$

which is plotted in Figure 5.9. Because all systems, regardless of their initial conditions or forcing function, can be cast in this form, we can refer to the function in Figure 5.9 as a *master curve* for under-damped second order systems.

The difference between the current system response variable, $\psi(t)$, and its value in steady state is the driver causing the dynamic response. In dimensionless form, $\hat{\psi}(t)$ represents the ratio of the current driving agent to the initial driving agent. The master curves are representative for any initial displacement, any set of system parameters for which the system remains underdamped, and any step input force. This master representation shows explicitly that with a sufficient

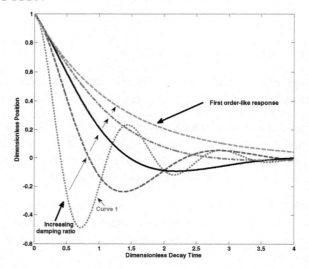

Figure 5.9: The response of a lumped mass with parallel spring and damper to a step input force from rest. The curves that are distinct in Figure 5.5 all collapse to the same curve (Curve 1). The rest of the curves correspond to increasing amounts of viscous damping in the system.

amount of damping, any second order system's response will be nearly indistinguishable from a corresponding first order-like response.

Further, since there are two initial conditions necessary for second order systems, there is a natural scaling of the initial velocity with $v^* = \zeta \omega_N (x_0 - x_{SS})$ such that the response is characterized by a dimensionless form of the initial velocity:

$$\hat{v}_0 = v_0/v^* = v_0/\zeta \omega_N (x_0 - x_{SS}).$$

The response of the original system is plotted in dimensionless form for a variety of initial velocities in Figure 5.10.

Ramp Input to an Over-damped System

We can maintain that the generalization holds when the system is exposed to a time-dependent forcing function. Consider the ramp input signal:

$$F_0(t) = \begin{cases} 0 & t < 0 \\ Dt & t \geq 0 \end{cases}.$$

The most general form of a linear function is then presumed for the particular solution:

$$\psi_p(t) = K + Ct.$$

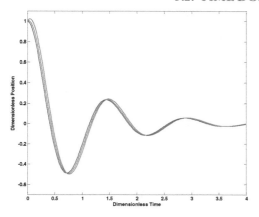 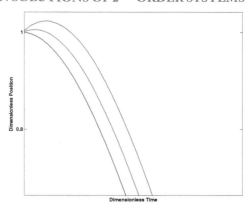

Figure 5.10: The response of a lumped mass with parallel spring and damper to a step input force for a variety of initial velocities.

Substituting this into the differential equation and setting like terms equal to one another:

$$\frac{1}{\omega_N^2}(0) + \frac{2\zeta}{\omega_N}C + K + Ct = \frac{Dt}{k}$$

$$\Rightarrow C = D/k$$

$$0 = K + \frac{2\zeta}{\omega_N}C$$

$$K = -\left(2\zeta/\omega_N\right)\frac{D}{k}$$

or

$$\psi_p(t) = -\frac{2\zeta}{\omega_N}\frac{D}{k} + \frac{D}{k}t = \frac{D}{k}\left(t - \frac{2\zeta}{\omega_N}\right).$$

Compiling the total solution and applying the initial conditions:

$$x(t) = x_h(t) + x_p(t)$$

$$= C_1 e^{\left(-\zeta\omega_N + \omega_N\sqrt{\zeta^2 - 1}\right)t} + C_2 e^{\left(-\zeta\omega_N - \omega_N\sqrt{\zeta^2 - 1}\right)t} + \frac{D}{k}\left(t - (2\zeta/\omega_N)\right).$$

Applying initial conditions:

$$x(0) = x_0 \Rightarrow x_0 = C_1 + C_2 - 2D\zeta/k\omega_N$$

$$\dot{x}(0) = x_0 \Rightarrow x_0 = C_1\left(-\zeta\omega_N + \omega_N\sqrt{\zeta^2 - 1}\right) + C_2\left(-\zeta\omega_N - \omega_N\sqrt{\zeta^2 - 1}\right) + \frac{D}{k}.$$

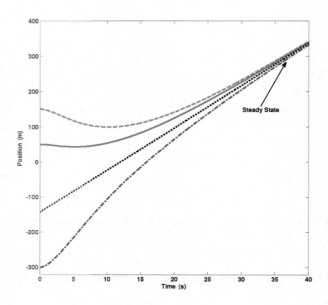

Figure 5.11: The response of an over-damped mass-spring-damper system to a ramp input force in dimensional form. The response is characterized by an exponential approach or transient from an initial value at rest to a final value that, like the forcing function, increases linearly in time.

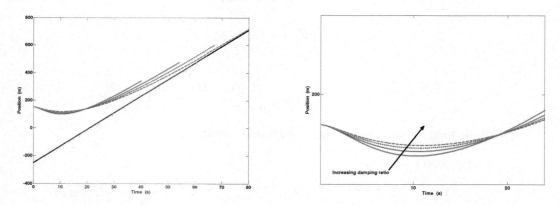

Figure 5.12: The response of an overdamped mass-spring-damper system to a ramp input force in dimensional form. With increasing values for the damping ratio, the response eventually appears first-order-like.

After resolving the values of C_1 and C_2, we plot the total response in Figure 5.3 for several distinct initial displacements. As for the under-damped case, systems with increasing damping ratios eventually respond in a manner that "looks" first-order-like (see Figure 5.12). Finally, solving the constants C_1 and C_2 for various initial velocities gives the responses shown in Figure 5.13.

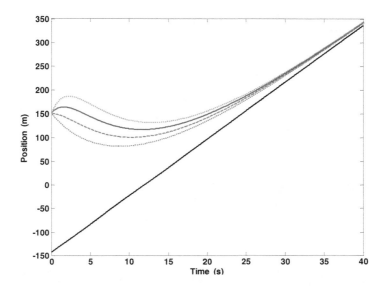

Figure 5.13: The response of an overdamped mass-spring-damper system to a step input force for various values of the initial velocity.

5.2.3 DIMENSIONLESS SOLUTIONS FOR 2nd ORDER SYSTEMS

Again, even when the steady state is time-dependent, the entire response can still, for *every* second order system, be cast in dimensionless form:

$$\hat{\psi}(t) = \frac{\psi(t) - \psi_{SS}(t)}{\psi_0 - \psi_{SS}(0)} = \Im\left(\hat{t}_1, \hat{t}_2\right) = \begin{cases} e^{-\hat{t}_1}\left[A\cos\left(2\pi\hat{t}_2\right) + B\sin\left(2\pi\hat{t}_2\right)\right] & \zeta < 1 \\ Ae^{\hat{t}} + B\hat{t}e^{\hat{t}} & \zeta = 1 \\ Ae^{-\hat{t}_1} + Be^{-\hat{t}_2} & \zeta > 1 \end{cases}$$

where the respective characteristic times are given in Table 5.6.

These solutions for the under and overdamped systems, respectively (in Section 5.2.2) are plotted in dimensionless form in Figure 5.14. Note the limit behaviors of under-damped and damped systems.

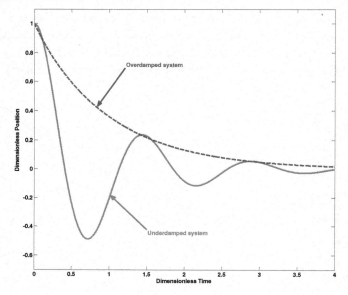

Figure 5.14: The response of a mass-spring-damper to an arbitrary input force in dimensionless form.

5.2.4 CHARACTERISTIC TIMES FOR TRANSIENTS IN 2nd ORDER SYSTEMS

The characteristic time for transients in any first order system corresponds directly with its system parameter, τ. Alternatively, the characteristic times associated with the transient response in 2nd order systems are functions of its system parameters as outlined in Table 5.6.

5.2.5 UNIVERSAL TRUTHS FOR 2nd ORDER SYSTEM RESPONSE IN THE TIME DOMAIN

We can now add several observations to our list of universal truths that always characterize how 2nd order systems respond to their environment. We note that 2nd order systems always approach a steady-state response from their initial state, and the response overshoots this steady response for under-damped systems and does not overshoot for over-damped systems. The steady response behaves like "a fence" that bounds the total response only when the system is over-damped. This over-damped response approaches the steady solution "from one side" as observed in Figure 5.11 for the ramp input example. We also note that even when the steady-state solution is time-dependent, the appropriate non-dimensionalization delivers a master curve that is identical for all initial conditions, i.e., starting points, and steady-state solutions, i.e., ending points.

Table 5.6: Characteristic times for transient solutions of second order systems

Scenario	ζ	Physical Regime	Characteristic Times
1	$\zeta < 1$	Under Damped	$\dfrac{1}{\zeta\omega_N}$ exponential decay $\dfrac{2\pi}{\omega_N\sqrt{1-\zeta^2}}$ damped period
2	$\zeta = 1$	Critically Damped	$\dfrac{1}{\zeta\omega_N}$ exponential decay
3	$\zeta > 1$	Over Damped	$\dfrac{1}{\left(\zeta\omega_N + \omega_N\sqrt{\zeta^2-1}\right)}$ first exponential decay $\dfrac{1}{\left(+\zeta\omega_N - \omega_N\sqrt{\zeta^2-1}\right)}$ second exponential decay

Of Special Note
   Universal Truths for 2nd Order Systems

(a) They are comprised of system elements (or characters) that store **both** potential **and** kinetic forms of energy

(b) Their behavior is characterized by a pair of system parameters, $\{\omega_N, \zeta\}$, where

(c) $\omega_N = f_1(L,C) = f_2(m,k)$ and $\zeta = g_1(L,C,R) = g_2(m,k,b)$

(d) The system transients are identified by two characteristic times

(e) The system, when underdamped, is capable of overshooting the eventual steady-state response.

(f) With an appropriate amount of damping, the system response is nearly indistinguishable from that of an appropriately parameterized first order system

(g) The system response can be universally placed in dimensionless form, normalized by two characteristic times.

5.2.6   ENERGY STORAGE AND DISSIPATION FOR 2nd ORDER SYSTEM RESPONSE IN THE TIME DOMAIN

Let's continue with the example of the mass-spring-damper system. The system stores both kinetic and potential energy. Now that we have resolved the resultant motion and velocity of the lumped mass analytically in Section 5.2.2, we may compute the energy partition that results from an imposed step input force applied to the mass when the system is underdamped (Figure 5.15).

The early transient behavior shows clearly that peak potential energy caches coincide with the absence of kinetic energy when the mass is at rest at peak values of displacement as shown in Figure 5.16. Behavior in the steady state shows the continued decay to a state of steady potential energy corresponding to the spring extended to its static deflection where motion ceases and kinetic energy decays to zero as shown in Figure 5.17. All the energy is eventually stored in the spring as the displacement converges on the static value. All the while, an order of magnitude more energy is dissipated in the damper throughout the transient as evidenced in Figure 5.18.

Note that the dissipated energy only ever increases. The work done by friction, as plotted in Figure 5.18, can never decrease and only ever accumulates.

This is perhaps more evident in an under-damped system that is given an initial displacement and released from rest. Here the entire response is simply a transient decay from the initial conditions. Recall from our discussion in Chapter 3 that in this case of a damped harmonic oscillator, the kinetic and potential caches are passed back and forth to one another while friction eats away during each transfer as shown in Figure 5.19. The energy story for each of the three characters (inertia, stiffness, and friction) is shown for a typical case in Figure 5.20.

In the resulting free response, energy is "consumed" within each exchange from kinetic to potential and back to kinetic. With each "pass of the energy ball" the total amplitude of stored energy is decreased by precisely the amount eaten away by friction as shown in Figure 5.21. Negligible energy is dissipated as the potential energy peaks, i.e., where the kinetic energy (and, therefore, velocity) is minimal. Most of the energy is dissipated where the kinetic energy (and velocity) reach their respective maxima.

Finally, consider the case of the over-damped system subjected to a ramp input. We solved the inertial displacement and velocity in Section 5.2.2. Here, owing to the slope of the ramp input force, the net kinetic energy stored plateaus at a relatively small value while the spring continues to stretch storing the lion's share of the imparted energy as potential. The dissipated energy also accounts for a substantial energy cache. These are shown in the early transient in Figure 5.19. Later, in the steady state the displacement becomes linear in time resulting in a potential energy cache that accumulates quadratically in time. The friction work is the integral of an F-v curve in the damper when the force approaches a constant value. In this case, the friction work increases linearly over long times. The stored kinetic energy plateaus along with the velocity at long times. Here, we recognize features of the solution without showing its explicit functional form. As Feynman correctly noted, "(We can) understand what an equation means if (we) have a way to figure out the characteristics of its solution without solving it."

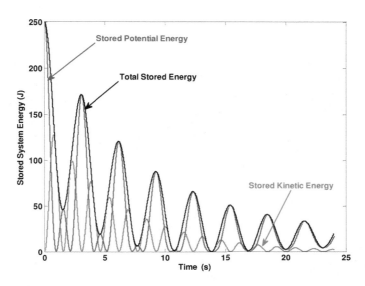

Figure 5.15: Energy partition for a mass-spring-damper system subject to a step input force of 60 N ($x_0 = 2$ m; $v_0 = 0$ m/s; $k = 125$ N/m; $b = 5$ Ns/m; $m = 30$ kg).

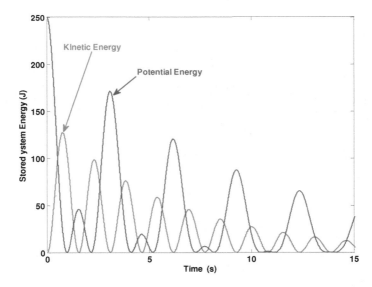

Figure 5.16: Energy partition for a mass-spring-damper system subject to a step input force of 60 N ($x_0 = 2$ m; $v_0 = 0$ m/s; $k = 125$ N/m; $b = 5$ Ns/m; $m = 30$ kg).

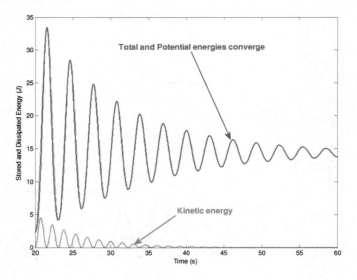

Figure 5.17: Steady-state energy partition for a mass-spring-damper system subject to a step input force of 60 N ($x_0 = 2$ m; $v_0 = 0$ m/s; $k = 125$ N/m; $b = 5$ Ns/m; $m = 30$ kg).

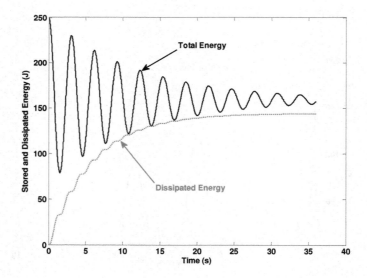

Figure 5.18: Total energy and dissipated energy for a mass-spring-damper system subject to a step input force.

Figure 5.19: The second order free response is the story of an energy catch between Captains Potential and Kinetic Energy while the Evil Dr. Friction "steals away" energy with each transfer.

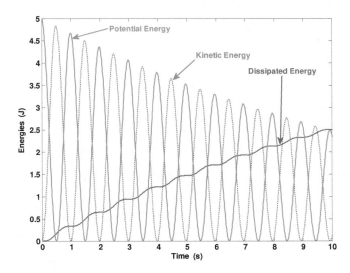

Figure 5.20: A second order system with dissipation is excited by an initial displacement from rest with no external forces applied.

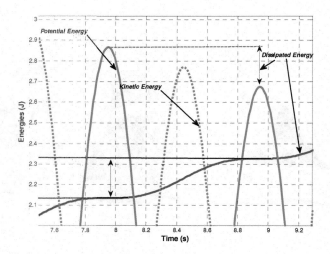

Figure 5.21: In the free response of an under-damped second order system with dissipation, with each "pass of the energy ball" the total amplitude of stored energy is decreased by precisely the amount eaten away by friction.

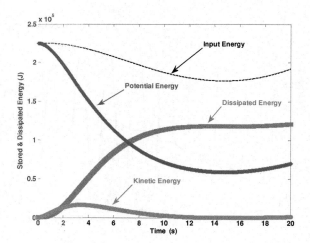

Figure 5.22: An over-damped second order system with ramp input experiences a continual introduction of energy to the system.

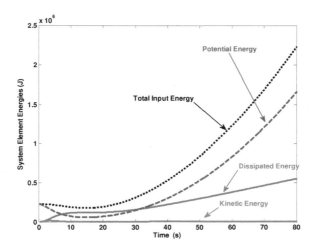

Figure 5.23: An over-damped second order system with ramp input as it enters steady state. Here, one can reason the forms of the steady dependence of energy dissipation (linear) and storage (quadratic) without actually solving the explicit equations.

5.3   CHAPTER ACTIVITIES

Problem 1 Consider the plate damper, mechanical system shown:

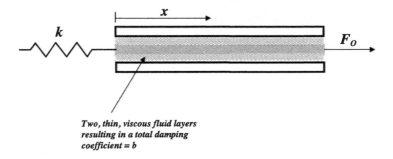

Two, thin, viscous fluid layers resulting in a total damping coefficient = b

If the mass is initially moving *to the right* with a velocity of 1 m/s from the position $x_0 = -2$ m and a constant, horizontal force is suddenly applied to the mass, as shown, write the differential equation governing the system plate displacement. What are the system's natural frequency and damping ratio? Sketch the system position response as a function of time. Be sure to specifically label initial conditions, steady-state response, transient response, and the settling time with numerical values where possible. Use $m = 0.1\,\text{kg}; k = 40\,\text{N/m}; b = 6\,\text{Ns/m}$.

Plot time histories for the system potential and kinetic energy caches as well as the energy dissipated over time.

Problem 2 Consider the plate damper, mechanical system from which the spring has been removed. The system is turned vertically and subject to a step input gravitational force as shown:

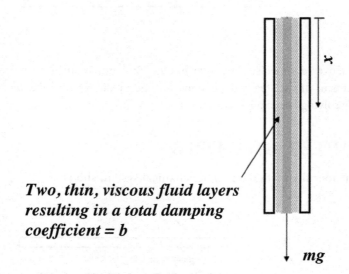

Two, thin, viscous fluid layers resulting in a total damping coefficient = b

If the mass is dropped from the position $x_0 = 0\,\text{m}$ from rest, write the differential equation governing the system plate velocity. Sketch the system response as a function of time. Be sure to specifically label initial conditions, steady-state response, transient response, and the settling time with numerical values where possible. Use $m = 4\,\text{kg}; b = 6\,\text{Ns/m}; g \approx 10\,\text{m/s}$.

Problem 3 Consider the mass-spring-damper system shown subject to a ramp input displacement of $y(t) = 5t$:

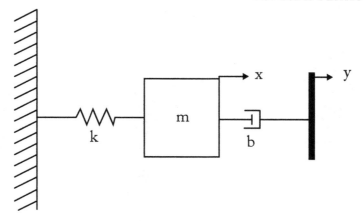

Derive the governing differential equation for the displacement of the mass. Solve the equation using $m = 10\,\text{kg}; k = 40\,\text{N/m}; b = 25\,\text{Ns/m}$. Plot the response, labeling the transient and steady regimes. Plot the displacement response in dimensionless form and compare with Figure 5.11.

Problem 4 Consider the downhill skier pictured here:

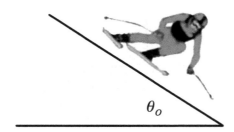

The total drag on the skier, F_D, is a combination of man-made-snow surface resistance and aerodynamic drag resulting in the following relationship for the drag force: $\vec{F_D} = C_D\,\vec{V}$ where C_D is the coefficient of drag, $\vec{V}$ is the velocity of the skier down the inclined slope and C_D = constant. Draw an appropriately labeled free body diagram and derive the equation governing the skier's velocity.

If the skier jumps out a gate and starts ideally from rest, determine:

(a) the skier's eventual terminal downhill velocity

(b) how long it will take to effectively attain this speed.

Use $m = 80\,\text{kg}; b = 16\,\text{Ns/m}; g \approx 10\,\text{m/s}$. Plot the energy stored and dissipated in the system over a relevant time scale. What story do they tell?

Problem 5 Consider the idealized windshield wiper mechanism illustrated here.

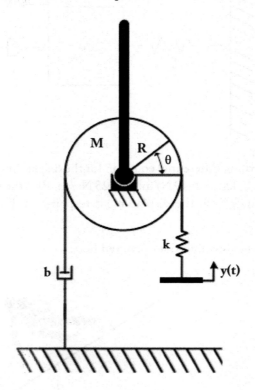

A mass-less blade is rigidly attached to the disk of radius R. Use $I \approx 1/2mR^2$ for the disk and wiper blade assembly for all calculations. Assuming the angular rotation of the disk remains "small," derive the differential equation governing the sweep of the wiper blade. Based on your differential equation, derive theoretical expressions for the system's natural frequency and damping ratio. What damping coefficient is required to critically damp the system?

Solve for the total response when the platform is subject to a step input displacement of $y(t) = 1.2$ inches using: $Y_{IN} = 1.2$ inches; $R = 0.5$ inches; $k = 1\,\text{lb/ft}$

$m = 0.01$ slug; $b = 0.25$ lb-s/ft.

Problem 6 Consider the angular position of a 100 kg winter Olympic snowboarder on a circular pipe of radius, R. The total drag on the snowboarder, F_D, is a combination of man-made-snow surface resistance and aerodynamic drag resulting in the following relationship for the drag force: $\vec{F}_D = C_D \vec{V}$ where C_D is the coefficient of drag and $\vec{V}$ is the tangential velocity of the snowboarder and $C_D = $ constant.

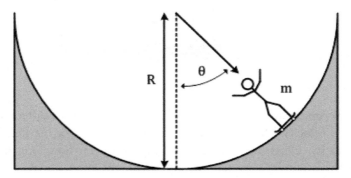

Use $I = mR^2$, $R = 10\,\text{m}$, $g = 10\,\text{m/s}^2$ for all calculations.

Assuming that the snowboarder enters the pipe at an initial position of $\theta = 30°$ and begins his angular descent from rest, show that the differential equation governing the angular position of our snowboarder with respect to time is given by

$$mR^2\ddot{\theta} + C_D R^2 \dot{\theta} + mgR \sin\theta = 0.$$

Consider that the small angle approximation is valid and that on two successive passes five seconds apart, the maximum angular values are:

$$\theta_N = 30° \quad \text{and}$$
$$\theta_{N+1} = 25°.$$

Using the log decrement, compute the system's natural frequency and damping ratio. Make a theoretically informed estimate of the drag coefficient, C_D, based on these measurements. From an initial angular entry point at $\theta_0 = 30°$, how long would it take the snowboarder to effectively come to rest? Present your solution in dimensionless form and compare a graph of the dimensionless position with Figure 5.11.

Problem 7 You're escaping the East India Trading Company in your trusty vessel "The Black Pearl." The Pearl's sails generate thrust in the following relationship:

$$F_{Sail} = C_S(V_W - V_P)$$

where V_P is the velocity of the Pearl, V_W is the velocity of the wind, and C_S is a constant. The drag on the Pearl's hull is linearly proportional to her velocity:

$$F_{Drag} = C_D V_P$$

where C_D and the Pearl's mass, m, are constant.

Use an appropriately labeled free body diagram to derive the differential equation governing the Pearl's velocity. Determine an algebraic expression for the Pearl's terminal, i.e., steady state, velocity. Determine an algebraic expression for how long it will take the Pearl to "effectively" attain its terminal velocity. Write out a functional solution for the velocity of the Pearl. Assume the initial velocity is given by V_{PO}. Sketch the solution for the Pearl's velocity. Identify the time constant, τ, and the corresponding terminal velocity, υ_{SS}, on the graph.

Problem 8 A pressure-compensating hydraulic spool valve consists of a bar-bell-like mass in a cylindrical sleeve (shown below). The valve is moved horizontally by a solenoid that applies a step input force to the mass. A spring at the far end provides an opposing force. Hydraulic fluid in a tight clearance of width, h, provides a viscous friction force resisting the motion and given by the relation:

$$F_\upsilon = \frac{C\upsilon}{h}$$

where C is a constant. A balance of forces in the horizontal direction gives:

$$m\frac{d^2x}{dt^2} = F(t) - kx - \frac{C\upsilon}{h}.$$

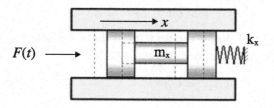

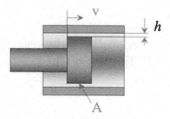

$$F(t) = \begin{cases} 0 & t < 0 \\ 1\,\text{N} & t \geq 0 \end{cases}$$

$$m = 0.01\,\text{kg}$$
$$k = 100\,\text{N/m}$$
$$h = 20 * 10^{-6}\,\text{m}$$

Upon step-input application of the solenoid force, the valve is designed to move horizontally *as fast as possible* to its equilibrium position without overshooting it and without oscillating.

(a) The governing equation $m\frac{d^2x}{dt^2} = F(t) - kx - \frac{Cv}{h}$ physically represents a statement of what balance principle?

(b) What value of C must be used for the steady-state amount of valve travel to be achieved in the minimum time without oscillation?

(c) What is the steady-state amount of horizontal travel realized by the valve under this step input force?

(d) Roughly how long will it take for the valve to travel to its equilibrium position?

(e) Plot the system's total energy stored and dissipated over time.

(f) Often hydraulic fluid becomes contaminated as wear particles accumulate in the clearance between the spool and its housing. Such particles often jam in the clearance effectively reducing the clearance width. Using arguments *supported by the form of the solution for the valve motion*, explain the effect the particulate contamination will have on the time necessary to move the valve to its steady-state position.

(g) If the value of the oil drag coefficient, C, used in part (a), were reduced to half its original value, would the system overshoot and oscillate about its eventual steady state? If so, with what frequency would it do so?

Problem 9 Consider the circuit shown with parallel system capacitors. At $t = 0$, a step voltage, V_0, is applied to the circuit by connecting it suddenly across a battery:

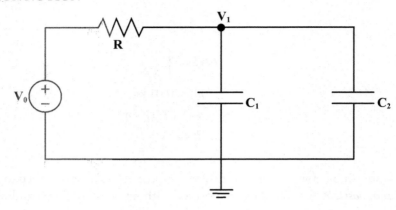

$$V_0 = 12\,\text{V}$$
$$i_R\,(t = 0) = 40\,milliamps$$

On the circuit diagram label the relevant nodes and apply the necessary conservation principles to derive the differential equation governing the response of the voltage drop across the pair of capacitors in the circuit. Use the potential energy storage system element equation to find the relevant initial condition or initial conditions for the system effort variable.

Sketch the system response as a function of time, labeling the output variable (on vertical axis), and the transient and steady-state regimes of behavior using $R = 100\,\Omega$; $C_1 = 25\,\mu f$; $C_2 = 100\,\mu f$.

Problem 10 Consider the system presented below in which the cord is wrapped around a solid disc with mass moment of inertia, $\frac{1}{2}MR^2$. The cord sticks to the disc without slipping. The disc is subjected to a ramp input torque, $T(t) = At$, applied about the fixed pivot at its center. The disc starts from rest at $\theta(0) = 0$ rad. Assume the disk rotation remains "small" and use an appropriately labeled free body diagram to derive the differential equation governing the disc's angular position, $\theta(t)$. Solve for the functional form of the disc position. At what time will the assumption of "small" angles break down? Assume angles of 30 degrees or less are reasonably "small." Express your answer in terms of $M, R, A, k_1,$ and k_2.

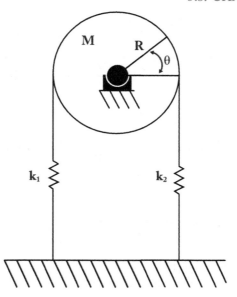

Problem 11 Consider the situation of drug absorption into a human being. The human body is your system and a drug is administered by the outside world at a rate given by $f(t)$. For such a case, the differential equation governing the amount of medicine in the blood stream, m, is given by:

$$\frac{dm}{dt} + rm = f(t) \qquad t \text{ in hours}$$

where $r = 0.0833\,\text{hr}^{-1}$.

The drugs are to be administered by injection which may be modeled as a non-zero initial condition: $f(t) = 0$, and $m\,(t = 0) = m_0 = 7\,\text{mg}$.

(a) Compute the solution for the presence of drug in the body over a representative time scale.

(b) What is the settling time for the drug to wear off?

(c) How many drug storage agent types are present in the system? Why?

(d) How many drug dissipation agent types are present in the system? Why?

Problem 12 Consider the mechanical system of the idealized building model below:

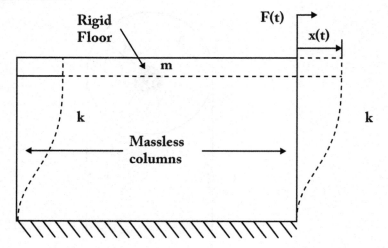

Take $m = 0.5$ slug; $k = 8$ lb/ft; $b = 1$ lb $-$ s/ft; $F(t) = F_0 = 32$ lb. If the mass is initially at rest at the position $x(0) = 0$ ft and a constant, horizontal 32 lb. force is suddenly applied to the right, as shown

(a) What are the system's natural frequency and damping ratio?

(b) Sketch the system response as a function of time. Be sure to specifically label initial conditions, steady-state response, transient response, and the settling time with numerical values where possible.

(c) What internal damping coefficient would be needed in the column walls to "just" make the building's lateral motion response behave "1st-order-like"?

CHAPTER 6

Going Nowhere?

Going from home to work to home to work, I am moving, but in the end I haven't gone anywhere ... vibrating strings move but go nowhere ... drawers open, close, open, close—all that motion and nothing to show for it. Oscillatory motion is interesting. Doing the same thing over and over and going nowhere is pretty important.

The Physics Hypertext Book

The conversion of circular motion into sine waves is a pervasive part of our daily lives. Sine waves are the atoms of structure. They're nature's building blocks. Primordial sine waves spanned the stuff of the cosmos. The ripples of a pond and the ridges of sand dunes are manifestations of the emergence of sinusoidal structure from a background of bland uniformity. There's something almost spiritual about them.

Steven Strogatz
The Joy of X

We've examined polynomial functions as input signals to dynamic systems. The category of harmonic functions is a special class unto itself and deserves individual treatment. Going to work and returning home, swinging on a swing in a playground, rotating a drum in a washing machine, spinning tires on an automobile—all are pervasive manifestations of periodicity in the world around us. And while one can admit the nature of periodicity is that one "goes nowhere," the energy story tells us something different. There is "something to show for it" in the energy tale. Response of a building to earthquake loading easily reminds us that only in one peculiar sense does the building "go nowhere." The ability of the building to absorb, store, or dissipate the input energy convinces us there is another side of the story.

There are myriad examples of periodic input that excite dynamics. Because the periodicity appears in the forcing function or excitation, we are interested in the steady-state solution long after the transient has decayed away. Normally such treatments are referred to as the frequency response of systems because the response is dependent on the frequency of the input excitation relative to the system. These solutions naturally appear in terms of the system parameters, where the specific mathematical form of the system parameters arises from the individual movie script.

6.1 FREQUENCY DOMAIN SOLUTIONS OF 1st ORDER SYSTEMS

First order systems result when the script involves a single *type* of storage element or character: either Captain Potential Energy or Captain Kinetic Energy. There may be multiple storage elements, but they must store only one type of energy. There can only be one storage superhero. In these cases, the governing differential equation has the form:

$$\tau \frac{d\psi}{dt} + \psi = \Psi_0(t) = \Psi_{IN}(t) = \Psi_{IN}\cos(\omega t).$$

In the time domain, we used the superposition of homogeneous and complimentary solutions to determine a total solution composed of both transient and steady state. For the unique case of periodic loading, the input excitation "never goes away." Therefore, one must be cognizant of the nature of the steady-state solution because it is the specific response to this ever-present input. The nature of the steady-state response to periodic input is captured in three characteristics:

(a) the solution to a periodic excitation of frequency, ω, is also a periodic function with the same frequency, ω

(b) the magnitude of the steady-state solution is a scale multiple of the input magnitude of the excitation and

(c) the solution is shifted in time from the input signal.

As such, the steady-state solution is always of the form:

$$\psi_{SS}(t) = \Psi_{OUT}\cos(\omega t + \varphi)$$

and we need only determine the magnitude, $\Psi_{OUT} = A\Psi_{IN}$, and phase shift, φ, in order to completely determine the periodic steady-state response of the system.

6.1.1 TRANSFER FUNCTION ANALYSIS FOR HARMONIC INPUT

Consider the case where the magnitude of the excitation, Ψ_{IN}, is constant, i.e., it is not a function of the excitation frequency. Because the steady state has *no memory* of the system's initial conditions, we assume zero initial conditions and apply the Laplace operator:

$$\mathcal{L}\left\{\tau \frac{d\psi(t)}{dt} + \psi(t) = \Psi_0(t) = \Psi_{IN}(t) = \Psi_{IN}\cos(\omega t)\right\} =$$

$$\tau s \Psi(s) + \Psi(s) = \Psi_{IN}(s)$$

$$(\tau s + 1)\Psi(s) = \Psi_{IN}(s).$$

Since $\Psi(s)$ represents the total output of the system when subject to input $\Psi_{IN}(s)$, we do not lose any generality by referring to it as $\Psi_{OUT}(s)$ giving

$$(\tau s + 1)\Psi_{OUT}(s) = \Psi_{IN}(s) \Rightarrow$$

$$G(s) = \frac{\Psi_{OUT}(s)}{\Psi_{IN}(s)} = \frac{1}{\tau s + 1}.$$

The parameter s is a quantity in the complex plane where $s = a + j\omega$. In the case where the time domain function is periodic and representable by trigonometric functions, the real portion of s dictates the exponential rate of decay for constant magnitude input. Since there is no decay in a pure sinusoid, in our case $a = 0$. The complex part remaining for the steady state is then simply $s = j\omega$. Using this simplification, we arrive at what is often called the sinusoidal transfer function (STF). For the remainder of this chapter, we will confine our discussions to STF's only. Making this substitution:

$$G(s = j\omega) = \frac{\Psi_{OUT}(j\omega)}{\Psi_{IN}(j\omega)} = \frac{1}{1 + \tau\omega j}.$$

Now the STF is a function whose numerator and denominator, in general, can be thought of as vectors in the complex plane

$$G(s = j\omega) = \frac{A + Bj}{C + Ej}$$

where $A = 1, B = 0, C = 1, E = \tau\omega$ for a first order system subject to constant magnitude periodic input. The STF can be used to easily compute the magnitude and phase shift of the resultant periodic response. The numerator and denominator vectors of the STF can be represented graphically in the complex plane (Figure 6.1), where:

$$G(s = j\omega) = \frac{A + Bj}{C + Ej} = \frac{\mathcal{N}}{\mathcal{D}}$$

$$\mathcal{N} = A + Bj = Ne^{j\alpha}$$

$$\mathcal{D} = C + Ej = De^{j\beta}$$

and

$$N = \sqrt{A^2 + B^2}$$

$$\alpha = \tan^{-1}(B/A)$$

$$D = \sqrt{C^2 + E^2}$$

$$\beta = \tan^{-1}(E/C).$$

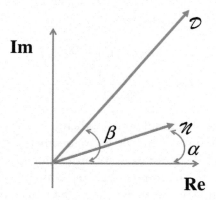

Figure 6.1: Graphical representation of the numerator and denominator vectors of the STF in the complex plane.

Now we may illustrate the utility of the Laplace approach for periodic input excitations. The STF, $G(j\omega)$ in this form can be used to readily obtain the magnitude and phase shift:

$$G(j\omega) = \frac{\Psi_{OUT}(j\omega)}{\Psi_{IN}(j\omega)} \Rightarrow \begin{cases} \dfrac{\Psi_{OUT}}{\Psi_{IN}} = \dfrac{\|\Psi_{OUT}(j\omega)\|}{\|\Psi_{IN}(j\omega)\|} = \dfrac{N}{D} = \dfrac{\sqrt{A^2 + B^2}}{\sqrt{C^2 + E^2}} = \mathcal{A} \\[2ex] \varphi = \angle\Psi_{OUT}(j\omega) - \angle\Psi_{IN}(j\omega) = \angle\mathcal{N} - \angle\mathcal{D} = \alpha - \beta \end{cases}$$

From this result:

$$\psi_{SS}(t) = \Psi_{OUT}\cos(\omega t + \varphi) = \mathcal{A}\Psi_{IN}\cos(\omega t + \varphi)$$

where, $\mathcal{A}$, the amplification ratio, and φ, the phase shift, of the response relative to the input excitation are given above as functions of the excitation frequency, ω, and the system parameters (here, τ).

6.1.2 STEADY-STATE RESPONSE AND BODE PLOT ANALYSIS

Frequency response is entirely characterized by the degree to which the output response is amplified and the degree to which the output response lags the input signal. Let's examine how this plays out for the simple case of the series RC circuit. Recall this circuit in Figure 6.2.

Consider the case where the input battery voltage or effort differential placed on the system is periodic with input magnitude, $V_{IN} = $ constant, such that

$$V_0(t) = V_{IN}\cos(\omega t)$$

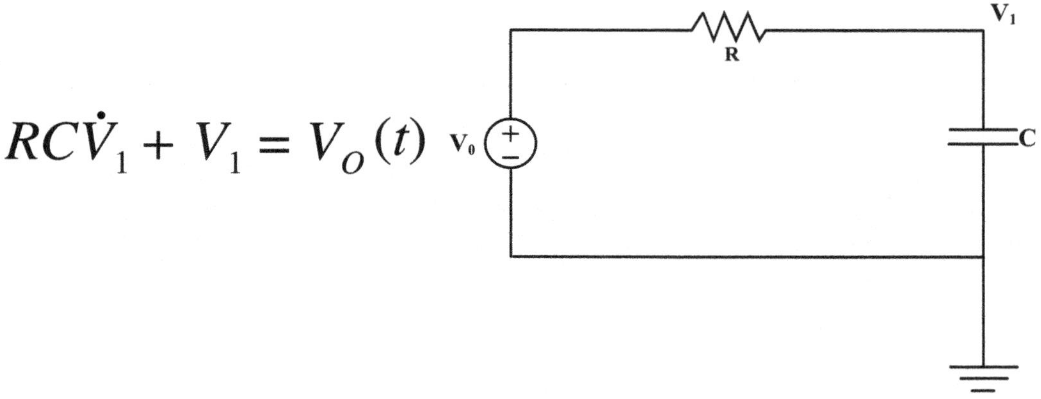

$$RC\dot{V}_1 + V_1 = V_O(t)$$

Figure 6.2: The series RC circuit and its governing differential equation.

where the magnitude $V_{IN} \neq f(\omega)$. Then, following the development of Section 6.1.1,

$$\mathcal{L}\left\{\tau\frac{dV_1(t)}{dt} + V_1(t) = V_{IN}\cos(\omega t)\right\} =$$
$$\tau s V_1(s) + V_1(s) = V_{IN}(s) \quad \Rightarrow$$
$$(\tau s + 1)V_1(s) = V_{IN}(s).$$

Since $V_1(s)$ represents the output voltage of the system, we can refer to it as $V_{OUT}(s)$ giving

$$(\tau s + 1)V_{OUT}(s) = V_{IN}(s)$$
$$G(s) = \frac{V_{OUT}(s)}{V_{IN}(s)} = \frac{1}{\tau s + 1}$$

which results in a sinusoidal transfer function

$$G(s = j\omega) = \frac{V_{OUT}(j\omega)}{V_{IN}(j\omega)} = \frac{1}{1 + \tau\omega j}.$$

From here, it is straightforward to calculate the amplification ratio

$$A = \frac{\|V_{OUT}(j\omega)\|}{\|V_{IN}(j\omega)\|} = \frac{\sqrt{1^2 + 0^2}}{\sqrt{1^2 + (\tau\omega)^2}} = \frac{1}{\sqrt{1 + (\tau\omega)^2}}$$

and the phase shift

$$\varphi = \angle\mathcal{N}(j\omega) - \angle\mathcal{D}(j\omega) = 0 - \tan^{-1}(\tau\omega)$$

of the system response

$$V_{1SS}(t) = \mathcal{A}V_{IN}\cos(\omega t + \varphi) = \frac{V_{IN}}{\sqrt{1 + (\tau\omega)^2}}\cos\left(\omega t - \tan^{-1}(\tau\omega)\right).$$

Note that all characteristics of the steady solution are only functions of the dimensionless quantity, $\tau\omega$. Plots of the amplification ratio (or alternatively the output response magnitude) and the phase shift as functions of the dimensionless quantity, $\tau\omega$, are known as the Bode plots. These are shown in Figures 6.3 and 6.4, respectively, below.

6.1.3   AN INTERPRETATION OF DIMENSIONLESS FREQUENCY RATIO

Often Bode plots are presented simply as a function of the dimensionless parameter, $\tau\omega$, which is sometimes referred to as the dimensionless frequency ratio. Whenever dimensionless parameters appear in a model, such parameters can often be placed in the form of a ratio of two physical quantities at play in the model. Let's examine how one may ascribe a physical interpretation to this dimensionless frequency ratio.

Consider the dimensionless parameter written as a ratio

$$\tau\omega = \frac{\omega}{1/\tau} = \frac{input\ excitation\ frequency}{equivalent\ system\ frequency}.$$

The input signal excites the system at an imposed frequency, ω. Alternatively, the "outside world" bombards the system with an imposed effort or flow at a rate of $f = \omega/2\pi$ cycles of input per second. This excitation is characterized by a characteristic time called its period, $T = 1/f = 2\pi/\omega$. So we see that the frequency can be interpreted as the reciprocal of the characteristic time. The larger the input signal frequency, the smaller its characteristic time. A similar interpretation can be had for the system. Since the system is characterized by its time constant, one can understand the time constant to be a measure of the system's response time, the time it takes the system to respond to external stimuli.

Now the dimensionless parameter, $\tau\omega$, as written above can be physically interpreted as a dimensionless frequency ratio: the ratio of the input excitation frequency to the frequency with which the system can respond to any input. When the excitation frequency is large compared to the frequency to which the system is capable of responding, then the excitation frequency is termed "high" in this relative sense. When the equivalent system frequency is large compared to the frequency imposed on it by "the outside world," then the excitation frequency is considered "low." When the ratio is of order unity, the frequency can be termed "moderate."

Summarizing

$$\tau\omega = \frac{\omega}{1/\tau} = \begin{cases} \gg 1 & \Rightarrow\ high\ excitation\ frequency \\ \approx 1 & \Rightarrow\ moderate\ excitation\ frequency \\ \ll 1 & \Rightarrow\ low\ excitation\ frequency. \end{cases}$$

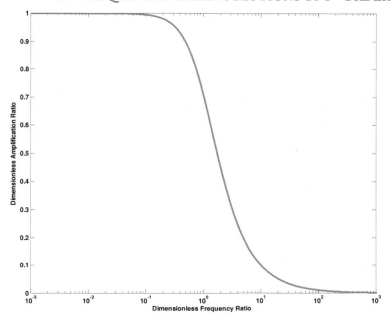

Figure 6.3: Amplification ratio as a function of dimensionless frequency ratio.

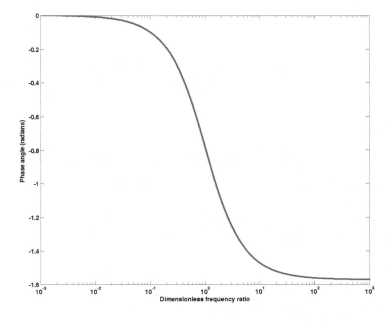

Figure 6.4: Phase shift as a function of dimensionless frequency ratio.

Similarly, consider the dimensionless parameter written as a ratio

$$\tau\omega = \frac{\tau}{1/\omega} = \frac{system\ characteristic\ time}{input\ excitation\ characteristic\ time}.$$

The system time constant is the characteristic response time of the system to external stimuli. The input signal bombards the system with an imposed effort or flow at a rate of $f = \omega/2\pi$ cycles of input per second or one dose every T seconds, where T is the input period, $T = 1/f = 2\pi/\omega$. If we consider the input excitation characteristic time to be a scaled quantity, $1/\omega$, we see that the excitation characteristic time can be interpreted as the reciprocal of the imposed frequency. The larger the input signal frequency, the smaller its characteristic time.

Now the dimensionless parameter, $\tau\omega$, as written above can be physically interpreted as a dimensionless characteristic response time ratio: the ratio of the time it takes the system to respond to an external stimulus to the characteristic time over which that stimulus is delivered by some external agent. When the system time constant is large compared to this characteristic time over which an excitation is delivered, the system is considered "slow to respond" or alternatively, the input is "coming at the system" faster than it can respond!

When the system time constant is small compared to this characteristic time over which an excitation is delivered, then the system response time is small relative to how often the stimulus is delivered. In this limit, the system is considered "fast to respond" or alternatively, the input is "coming at the system" slower than that rate at which the system can respond!

When the ratio is of order unity, the system can respond on time scales commensurate with those over which the excitation is being delivered.

Summarizing

$$\tau\omega = \frac{\tau}{1/\omega} = \begin{cases} \gg 1 \Rightarrow FAST\ system\ relative\ to\ the\ "outside\ world" \\ \approx 1 \Rightarrow system\ is\ of\ similar\ relative\ "speed"\ as\ the\ "outside\ world" \\ \ll 1 \Rightarrow SLOW\ system\ relative\ to\ the\ "outside\ world" \end{cases}.$$

These interpretations are summarized in Table 6.1.

6.1.4   FILTERING CHARACTERISTICS OF 1st ORDER SYSTEMS

In the classic sense of a frequency response, Bode plots show an infinite number of potential steady-state solutions each at a different imposed excitation frequency. The plots, because they are characterized by the dimensionless parameter, $\tau\omega$, exhibit unique behavior in the relatively low, moderate, and high frequency regimes.

Low Pass Filters
For the series RC circuit, the Bode plots are illustrated in Figures 6.3 and 6.4. In the low frequency regime, the amplitude ratio approaches unity and the output is negligibly shifted in time. In other words, the magnitude of the output voltage across the capacitor is nearly the same value as that

Table 6.1: Physical interpretations of the dimensionless frequency ratio, $\tau\omega$

Dimensionless Frequency Ratio	High Input Excitation Frequency	Low Input Excitation Frequency
$\tau\omega = \dfrac{\omega}{1/\tau}$	$\omega \gg \dfrac{1}{\tau}$	$\omega \ll \dfrac{1}{\tau}$
Dimensionless Characteristic Time Ratio	Fast System Response	Slow System Response
$\tau\omega = \dfrac{\tau}{1/\omega}$	$\tau \ll \dfrac{1}{\omega}$	$\tau \gg \dfrac{1}{\omega}$

input to the system by the external battery. In this limit, the steady-state output precisely mimics the input signal as shown in Figure 6.5.

For moderate excitation frequencies, the amplitude ratio approaches and $\sqrt{2}$ the phase shift approaches 45 degrees as shown in Figure 6.6.

In the high frequency regime, the amplitude ratio approaches zero and the phase shift approaches 90 degrees making the output a sine wave response to a cosine input. The output has negligible magnitude and lags the input signal as much as possible as shown in Figure 6.7.

The series RC circuit passes through all of the input excitation to the system at low input frequencies and passes none of the input signal and lags as much as possible at high input frequencies. For this reason the system is referred to as a *low pass filter*.

High Pass Filters

That the series RC circuit happened to behave as a low pass filter is entirely a result of its transfer function. It depends on both the nature of the excitation, the numerator in the transfer function, and the system itself, the denominator in the transfer the function. Change *either* the system, its elements or their structure *or* the nature of the input excitation and you *necessarily* change the transfer function, the representative Bode plots, and the filtering characteristics of the excited system.

So let's consider an alternate mechanical system with a mass-less platform sandwiched between a linear spring and damper as shown in Figure 6.8.

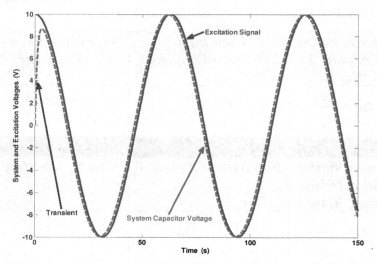

Figure 6.5: Series RC circuit response to low frequency excitation. This system is characterized by a time constant of 1 second and a transient of approximately 4 seconds after which time the response is predominantly steady state.

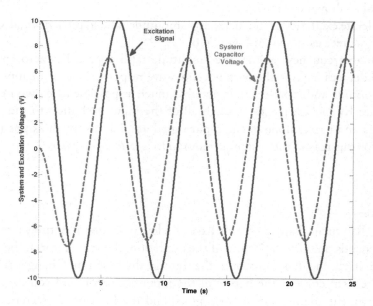

Figure 6.6: Series RC circuit response to moderate frequency excitation. Once again, the system settling time is roughly 4 seconds.

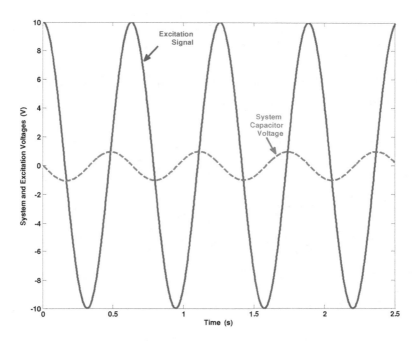

Figure 6.7: Series RC circuit response to high frequency excitation.

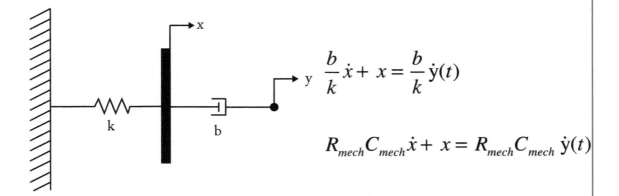

Figure 6.8: The mechanical spring–damper system with an interposed mass-less platform.

Operating on the governing differential equation with the Laplace operator

$$\mathcal{L}\left\{\tau\frac{dx(t)}{dt} + x(t) = \tau\frac{dy(t)}{dt}\right\}$$
$$= \tau s X(s) + X(s) = \tau s Y(s).$$

We understand the input to the system is the displacement of the end of the damper, $Y(s)$, and the output is the displacement of the mass-less platform, $X(s)$, giving

$$(\tau s + 1)X_{OUT}(s) = \tau s Y_{IN}(s)$$
$$G(s) = \frac{X_{OUT}(s)}{Y_{IN}(s)} = \frac{\tau s}{\tau s + 1}.$$

Calculating the amplification ratio

$$\mathcal{A} = \frac{\|X_{OUT}(j\omega)\|}{\|Y_{IN}(j\omega)\|} = \frac{\sqrt{0^2 + (\tau\omega)^2}}{\sqrt{1^2 + (\tau\omega)^2}} = \frac{\tau\omega}{\sqrt{1 + (\tau\omega)^2}}$$

and the phase shift

$$\varphi = \angle\mathcal{N}(j\omega) - \angle\mathcal{D}(j\omega) = \frac{\pi}{2} - \tan^{-1}(\tau\omega)$$

the system response is given by

$$X_{SS}(t) = \mathcal{A}Y_{IN}\cos(\omega t + \varphi) = \frac{\tau\omega Y_{IN}}{\sqrt{1 + (\tau\omega)^2}}\cos\left(\omega t - \frac{\pi}{2} + \tan^{-1}(\tau\omega)\right).$$

Again, all characteristics of the steady solution are only functions of the dimensionless quantity, $\tau\omega$. Plots of the amplification ratio and the phase shift are shown in Figures 6.9 and 6.10, respectively, below.

The mass-less platform exhibits quite different behavior. Here it is in the *high* frequency regime that the amplitude ratio approaches unity and the phase shift approaches zero degrees. In other words, the steady-state platform displacement precisely mimics the input signal as shown in Figure 6.11.

It is good to ask what is happening physically in this limit. When the right end of the damper is displaced at very high frequency, one is essentially applying a large periodic velocity here. When a large velocity differential is applied across a damper, it locks up and behaves as if it is rigid. The displacement time histories of both the input excitation and the platform motion should be identical in this limit.

Alternatively, when the damper's right end is harmonically displaced at extremely low frequency, it is the same as applying an infinitesimal velocity differential across the damper or negligible force. In this limit, the lion's share of the displacement across the damper occurs at the right

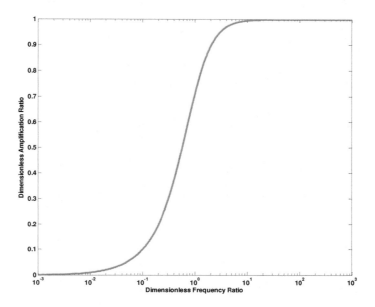

Figure 6.9: Amplification ratio as a function of dimensionless frequency ratio.

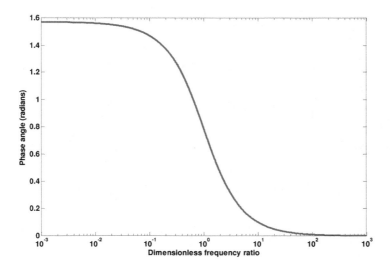

Figure 6.10: Phase shift as a function of dimensionless frequency ratio.

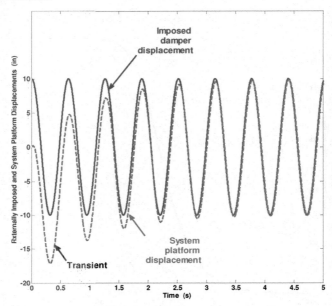

Figure 6.11: Mass-less platform response to high frequency excitation. The settling time for this system is approximately 2.5 seconds.

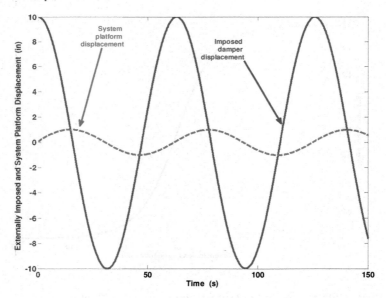

Figure 6.12: Mass-less platform response to low frequency excitation.

end while the magnitude of displacement of the system platform is negligible. Also the platform displacement lags the input displacement history by "as much as possible" or 90 degrees resulting in a platform displacement that is a sine wave response to a cosine input as shown in Figure 6.12.

Because this mechanical system passes all of the input excitation to the system at high input frequencies and passes none of the input signal and lags as much as possible at low input frequencies, the system is referred to as a *high pass filter*.

6.1.5 UNIVERSAL TRUTHS FOR 1st ORDER SYSTEMS SUBJECT TO HARMONIC INPUT

Of Special Note

For all first order systems, certain steady-state behaviors are characteristic of all systems:

(a) Their steady-state behavior is a function of a single dimensionless parameter, $\tau\omega$

(b) Dimensionless amplification ratios can never exceed a value of unity

(c) Phase shifts can never exceed 90 degrees

(d) Bode plots contain, at most, a single inflection point:

(a) Order 1 equation $\Rightarrow$ 1 inflection point

(b) One inflection point $\Rightarrow$ amplification ratio and phase monotonically either increase or decrease with $\tau\omega$.

(e) Systems may only ever be low-pass or high-pass filters.

6.1.6 ENERGY STORAGE AND DISSIPATION IN 1st ORDER SYSTEMS SUBJECT TO HARMONIC INPUT EXCITATION

Let's continue with the example of the mass-less spring-damper system discussed in Section 6.1.4. Because the idealized platform has negligible mass, no kinetic energy can be stored by the system. We know that energy can only be stored in the form of potential energy in the spring or dissipated by the damper. Now that we have resolved the resultant motion and velocity of the platform, we may compute the energy partition that results from an imposed harmonic input to the damper.

For zero initial conditions, the total platform displacement can be written as

$$x(t) = x_{TRANSIENT}(t) + x_{STEADY\ STATE}(t)$$
$$= (x_0 - x_{SS\_0})\,e^{-t/\tau} + AY_{IN}\cos(\omega t + \varphi)$$
$$= (x_0 - x_{SS\_0})\,e^{-t/\tau} + \frac{\tau\omega Y_{IN}}{\sqrt{1 + (\tau\omega)^2}}\cos\left(\omega t - \frac{\pi}{2} + \tan^{-1}(\tau\omega)\right).$$

The potential energy is simply

$$V_{SYSTEM}(t) = \frac{1}{2}kx^2.$$

While the energy dissipated in the damper is equal to the friction work performed by the damper

$$W_{FRICTION}(t) = \int_0^t F_{FRICTION}(t)dx$$
$$= \int_0^t b(\dot{y}(t) - \dot{x}(t))^2 dt.$$

These quantities are shown graphically in Figures 6.13 and 6.14, respectively.

The energy story tells an interesting tale that is potentially belied by the frequency response alone. At low input frequency, there is negligible movement of the platform. While the platform displacement is relatively low compared with the damper stroke displacement, it is not zero. As a result, the spring potential energy, is relatively speaking, low. The relative velocity over the damper, however, results in energy dissipation that dominates the energy story. It is nearly two orders of magnitude larger than the potential energy stored in the system.

At high frequency, the damper appears effectively locked, but there remains a relative velocity over the damper that can be relatively large owing to the high frequency of the damper stroke displacement. Therefore, the energy dissipated in the damper still dominates, only now it is only half an order of magnitude larger. The relative amount of energy stored has increased compared with the case at low frequency.

It is important that this result explicitly depends on the values of spring constant and damping coefficient and not simply their ratio, the time constant. Therefore, the energy story of two systems with the same time constant will not necessarily be the same as is the story for effort and flow. But the relative amounts of energy stored and spent will potentially be a deciding factor in system design.

This is an important issue not often discussed in elementary courses in systems dynamics. It plays a significant role in that while one needs to know the flow variables of velocity and displacement to calculate the kinetic and potential energies stored by the system, it may be the energy storage versus dissipation that is the deciding factor in the feasibility of the design. An analogous issue arises in finite element analysis where the primary solution variables are a set of nodal point displacements in a loaded structure. While this is true, it is often the internal stresses that

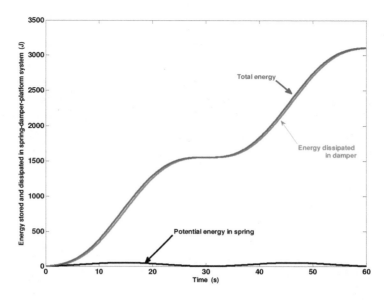

Figure 6.13: Energy partition for a mass-less platform response for low input frequency excitation.

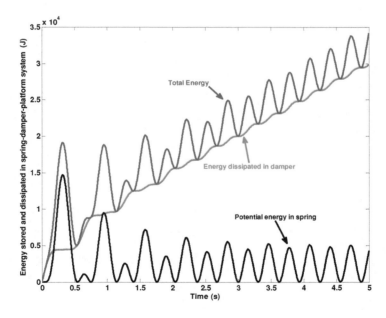

Figure 6.14: Energy partition for a mass-less platform response for high input frequency excitation.

are the determining factor in design. The internal stresses are calculated by using displacements to compute strains and strains to compute stresses. That is, the displacements or flow variables, by themselves, are incidental. The corresponding transmitted forces or effort variables internal to the system and energies stored are primary factors in system design. In engineering system design, engineers often redesign systems to lower transmitted forces or internally stored energy. This is accomplished by either altering the geometric structure of the system, i.e., whether system elements are connected in series or parallel, or by altering the material properties, i.e., the system capacitances, inductances, or resistances in any given geometric configuration. It's not unlike playing with Lego bricks. They can be put together in an infinite number of ways and we can choose different sizes of bricks. When the spring and damper are placed on either side of the platform and the outside world is stroking the damper, as shown here, the requisite energy losses are substantial. As we will see, such is not always the case.

6.2   FREQUENCY DOMAIN SOLUTIONS OF 2nd ORDER SYSTEMS

> If you wish to understand the universe, think of energy, frequency, and vibration.

> Nikola Tesla

Second order systems result when the script involves multiple *types* of storage elements or characters, i.e., Captains Potential and Kinetic Energy both appear in the movie. Note there may be many storage elements, but they must collectively store both types of energy. In these cases, the governing differential equation has the form:

$$\frac{1}{\omega_N^2}\frac{d^2\psi(t)}{dt^2} + \frac{2\zeta}{\omega_N}\frac{d\psi(t)}{dt} + \psi(t) = \Psi_0(t) = \Psi_{IN}(t) = \Psi_{IN}\cos(\omega t)$$

where ψ represents the pertinent effort or flow variable that characterizes the system and ω_N and ζ are the system natural frequency and damping ratio, respectively. Again, for periodic loading, the input excitation "never goes away." The steady-state solution is a response specifically to this omnipresent input excitation. Just as with 1st order systems, the nature of the steady-state response to periodic input is captured in three characteristics:

(a) the solution to a periodic excitation of frequency, ω, is also a periodic function with the same frequency, ω

(b) the magnitude of the steady-state solution is a scale multiple of the input magnitude of the excitation and

(c) the solution is shifted in time from the input signal

Therefore, the steady-state solution is always of the form:

$$\psi_{SS}(t) = \Psi_{OUT} \cos(\omega t + \varphi)$$

and we need only to determine the magnitude, $\Psi_{OUT} = A\Psi_{IN}$, and phase shift, φ, in order to completely determine the periodic steady-state response of the system. Because the steady-state solution has no memory of the system's initial conditions, we, again, use Laplace transforms to examine a system's steady-state response to periodic input.

6.2.1   TRANSFER FUNCTION ANALYSIS FOR HARMONIC INPUT

Consider the case where the magnitude of the excitation, Ψ_{IN}, is constant, i.e., it is not a function of the excitation frequency. Again, assuming zero initial conditions and applying the Laplace operator to the differential equation:

$$\mathcal{L}\left\{ \frac{1}{\omega_N^2} \frac{d^2\psi}{dt^2} + \frac{2\zeta}{\omega_N} \frac{d\psi}{dt} + \psi = \Psi_0(t) = \Psi_{IN}(t) = \Psi_{IN} \cos(\omega t) \right\} =$$

$$\frac{1}{\omega_N^2} s^2 \Psi(s) + \frac{2\zeta}{\omega_N} s\Psi(s) + \Psi(s) = \Psi_{IN}(s)$$

$$\left(\frac{1}{\omega_N^2} s^2 + \frac{2\zeta}{\omega_N} s + 1 \right) \Psi_{OUT}(s) = \Psi_{IN}(s)$$

$$G(s) = \frac{\Psi_{OUT}(s)}{\Psi_{IN}(s)} = \frac{1}{\left(\frac{1}{\omega_N^2} s^2 + \frac{2\zeta}{\omega_N} s + 1 \right)}.$$

For periodic input, $s = j\omega$, rendering the second order sinusoidal transfer function:

$$G(s = j\omega) = \frac{\Psi_{OUT}(j\omega)}{\Psi_{IN}(j\omega)} = \frac{1}{\left(1 - \frac{\omega^2}{\omega_N^2} \right) + \frac{2\zeta}{\omega_N} \omega j}.$$

Now

$$G(s = j\omega) = \frac{A + Bj}{C + Ej}$$

where $A = 1, B = 0, C = \left(1 - \frac{\omega^2}{\omega_N^2} \right), E = \frac{2\zeta}{\omega_N}$ for a second order system subject to constant magnitude periodic input.

The amplification ratio and phase shift follow:

$$G\left(j\omega\right) = \frac{\Psi_{OUT}(j\omega)}{\Psi_{IN}(j\omega)}$$

$$\Rightarrow \begin{cases} \dfrac{\Psi_{OUT}}{\Psi_{IN}} = \dfrac{\|\Psi_{OUT}\left(j\omega\right)\|}{\|\Psi_{IN}\left(j\omega\right)\|} = \dfrac{N}{D} = \dfrac{\sqrt{A^2 + B^2}}{\sqrt{C^2 + E^2}} = \mathcal{A} \\ \varphi = \angle\Psi_{OUT}\left(j\omega\right) - \angle\Psi_{IN}\left(j\omega\right) = \angle N - \angle D = \alpha - \beta \end{cases}$$

and

$$\psi_{SS}(t) = \Psi_{OUT}\cos(\omega t + \varphi) = \mathcal{A}\Psi_{IN}\cos(\omega t + \varphi)$$

where $\mathcal{A}$ and φ are functions of ω_N and ζ.

6.2.2   STEADY-STATE RESPONSE AND BODE PLOT ANALYSIS

For second order systems, the concept of a frequency ratio is explicit as the system is characterized by its natural frequency as opposed to a time parameter as in first order systems. Again, the specific instances of periodic signal inputs are best shown by specific examples.

Periodic Input Signal of Constant Magnitude

Consider the classical mass-spring-damper system from Section 4.4.1 and illustrated in Figure 4.12. Let's restrict ourselves to the case where the externally applied input force or effort placed on the system is periodic with input magnitude, $F_{IN} = $ constant, such that $F_0(t) = F_{IN}\cos\left(\omega t\right)$ where the magnitude $F_{IN} \neq f\left(\omega\right)$. Then, following the development of Section 6.1.1,

$$\left(\frac{1}{\omega_N^2}s^2 + \frac{2\zeta}{\omega_N}s + 1\right)X(s) = \frac{1}{k}F_{IN}(s)$$

$$G(s) = \frac{X(s)}{F_{IN}(s)/k} = \frac{1}{\left(\dfrac{1}{\omega_N^2}s^2 + \dfrac{2\zeta}{\omega_N}s + 1\right)}$$

which results in a sinusoidal transfer function

$$G\left(j\omega\right) = \frac{X_{OUT}(j\omega)}{F_{IN}(j\omega)/k} = \frac{1}{\left(1 - \dfrac{\omega^2}{\omega_N^2}\right) + \dfrac{2\zeta}{\omega_N}\omega j}.$$

The resulting amplification ratio is given by:

$$\mathcal{A} = \frac{\|X_{OUT}\left(j\omega\right)\|}{\|F_{IN}(j\omega)/k\|} = \frac{\sqrt{1^2 + 0^2}}{\sqrt{\left(1 - \dfrac{\omega^2}{\omega_N^2}\right)^2 + \left(2\zeta\dfrac{\omega}{\omega_N}\right)^2}} = \frac{1}{\sqrt{(1 - r^2)^2 + (2\zeta r)^2}}$$

where $r = \omega / \omega_N$ is known as the dimensionless frequency ratio. The corresponding phase shift is given by:

$$\varphi = \angle \mathcal{N}(j\omega) - \angle \mathcal{D}(j\omega) = 0 - \tan^{-1}\left(2\zeta r/(1 - r^2)\right).$$

So, finally, in steady state

$$X_{SS}(t) = \mathcal{A}\frac{F_{IN}}{k}\cos(\omega t + \varphi) = \frac{F_{IN}/k}{\sqrt{(1 - r^2)^2 + (2\zeta r)^2}}\cos\left(\omega t - \tan^{-1}\left(2\zeta r/(1 - r^2)\right)\right).$$

Note that all characteristics of the steady solution are only functions of the dimensionless quantities, ζ and r. Plots of the amplification ratio and the phase shift as functions of the dimensionless quantities, ζ and r, are known as the Bode plots or surfaces for second order systems. These are shown in Figures 6.15 and 6.16, respectively, for the case of constant magnitude input signal.

We should note several observations for this specific case of a constant force amplitude periodic signal input to a parallel mass-spring-damper system:

(a) at low frequency ratio, all of the signal input is passed onto the system with an amplification of zero and zero phase shift.

(b) At frequency ratios near unity, where the input signal frequency equals the system natural frequency, the amplification ratio can become much larger than one. For an undamped system, the output system response magnitude will grow unbounded at $r = 1$. This is known as *resonance*.

(c) At high frequency ratio, the amplification ratio falls off monotonically and asymptotically to zero at sufficiently high frequency ratio.

(d) The amplification ratio always decreases with increasing damping for all frequency ratios.

(e) At sufficiently high damping ratio, the system appears first-order-like and behaves like a low pass filter.

In most cases, one cannot make generalizations about the behavior of any one system from a different system. To see how any periodically excited system behaves in the steady state, one must derive the transfer function and examine the behavior in the Bode plots. The transfer function depends both on the system parameters and features of the forcing function. Whenever either is altered, the transfer function and steady-state behavior can be altered. Each system under specific signal inputs must be examined on its own merits. Considering a second example will make this point unambiguous.

Periodic Input Signal from a Rotating Imbalance

When rotating machinery is submitted to an imbalance about the axis of rotation, such as happens when wet clothes shift to one side of a spinning basin in a washing machine, the washing machine

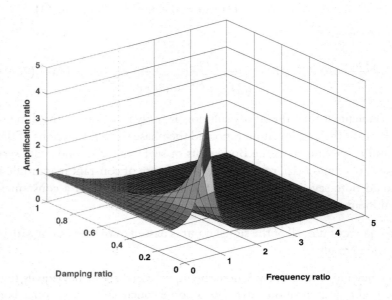

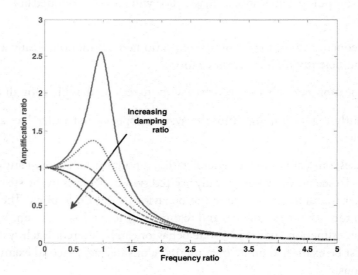

Figure 6.15: Amplification ratio as a function of frequency and damping ratios.

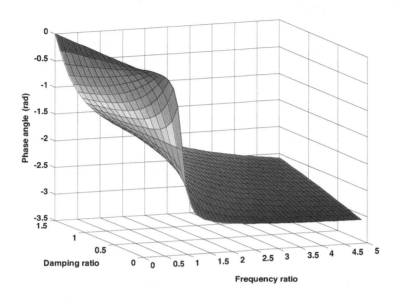

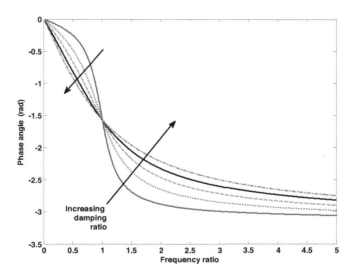

Figure 6.16: Phase shift as a function of frequency and damping ratios.

is excited into motion. Similarly, an automobile exhibiting a wheel imbalance will have observable and detrimental vibration imparted to the car axle and frame. A simple lumped model of such an inertial imbalance is illustrated in Figure 6.17.

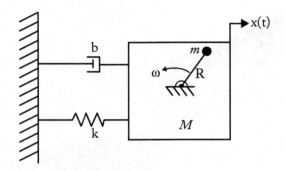

Figure 6.17: A second order system subject to a rotating imbalance.

The system is characterized by some frictional losses that we assume can be modeled effectively as viscous dissipation with damping coefficient, b, and a system stiffness, k, whereby the system stores potential energy. The relatively small imbalance ($m \ll M$) is spinning about a frame rigidly attached to the mass, M, at constant angular velocity, ω. The imbalance, m, is spinning at a prescribed rotational speed, thus imparting an eccentric load on the inertial mass, M, that is sinusoidal with a magnitude that is dependent on the spinning speed. Because the spinning speed is prescribed, the system has only a single degree of freedom. This is often called a classical rotating imbalance. Consider the location of the mass imbalance relative to the center of the lumped system mass to be given by a vector, $\mathcal{R}(t) = R \cos \omega t \hat{i} + R \sin \omega t \, \hat{j}$ where R is the magnitude of the eccentricity of the imbalance. If the block is constrained in the vertical ($\hat{j}$) direction, a free body diagram on the inertial block renders the following governing differential equation for the horizontal motion of the mass, M:

$$M \frac{d^2 x(t)}{dt^2} + b \frac{dx(t)}{dt} + k x(t) = -m \ddot{R}(t) = mR\omega^2 \cos(\omega t).$$

Normalizing the equation by the system stiffness, k, and assuming $M \gg m$,

$$\frac{M}{k} \frac{d^2 x(t)}{dt^2} + \frac{b}{k} \frac{dx(t)}{dt} + x(t) = \frac{mR\omega^2}{k} \cos(\omega t) = \frac{mR\omega^2}{M\omega_N^2} \cos(\omega t)$$

$$\frac{1}{\omega_N^2} \frac{d^2 x(t)}{dt^2} + \frac{2\zeta}{\omega_N} \frac{dx(t)}{dt} + x(t) = \frac{mR\omega^2}{M\omega_N^2} cos(\omega t).$$

Note here that the magnitude of the forcing function is dependent on the frequency of rotation of the imbalance. The magnitude of the imbalance increases as the square of the spinning frequency.

We will see that this is a crucial feature of this excitation. Applying the Laplace operator to the differential equation:

$$\mathcal{L}\left(\frac{M}{k}\frac{d^2x(t)}{dt^2} + \frac{b}{k}\frac{dx(t)}{dt} + x(t) = -\frac{m}{k}\ddot{R}(t)\right)$$

$$\left(\frac{1}{\omega_N^2}s^2 + \frac{2\zeta}{\omega_N}s + 1\right)X(s) = \frac{-m}{k}s^2R(s) = \frac{-M}{k}\frac{m}{M}s^2R(s) = \frac{-ms^2}{M\omega_N^2}R(s)$$

$$G(s) = \frac{X(s)}{R(s)} = \frac{-ms^2/M\omega_N^2}{\left(\dfrac{1}{\omega_N^2}s^2 + \dfrac{2\zeta}{\omega_N}s + 1\right)}$$

which results in a sinusoidal transfer function

$$G(j\omega) = \frac{X_{OUT}(j\omega)}{R_{IN}(j\omega)} = \frac{m\omega^2/M\omega_N^2}{\left(1 - \dfrac{\omega^2}{\omega_N^2}\right) + \dfrac{2\zeta}{\omega_N}\omega j}$$

and amplification ratio

$$\mathcal{A} = \frac{\|MX_{OUT}(j\omega)\|}{\|mR_{IN}(j\omega)\|} = \frac{\left(\omega^2/\omega_N^2\right)}{\sqrt{\left(1 - \dfrac{\omega^2}{\omega_N^2}\right)^2 + \left(2\zeta\dfrac{\omega}{\omega_N}\right)^2}} = \frac{r^2}{\sqrt{(1-r^2)^2 + (2\zeta r)^2}}$$

where $r = \omega/\omega_N$, and, again, $\mathcal{A} = \mathcal{A}(r, \zeta)$.

The corresponding phase shift is given by:

$$\varphi = \angle\mathcal{N}(j\omega) - \angle\mathcal{D}(j\omega) = 0 - \tan^{-1}\left(2\zeta r/\left(1 - r^2\right)\right).$$

So, finally

$$X_{SS}(t) = \mathcal{A}\frac{mR_{IN}}{M}\cos(\omega t + \varphi) = \frac{(mR_{IN}/M)r^2}{\sqrt{(1-r^2)^2 + (2\zeta r)^2}}\cos\left(\omega t - \tan^{-1}\left(2\zeta r/\left(1 - r^2\right)\right)\right).$$

While the phase shift is identical to that for the constant magnitude forcing function, the presence of r^2 in the numerator changes the amplification ratio in significant ways. The resultant Bode plot of the amplification ratio is shown in Figure 6.18.

For the specific case of a periodic signal input from a rotating imbalance to a parallel spring–damper—mass system:

(a) at low frequency ratio, none of the signal input is passed onto the system with an amplification of zero and zero phase shift.

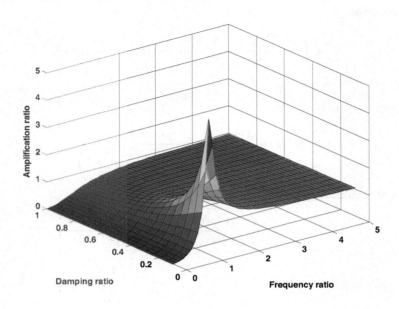

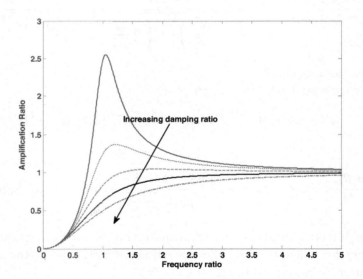

Figure 6.18: Amplification ratio for a rotating imbalance.

(b) At near resonant frequencies, the amplification ratio can become much larger than one. For an undamped system, the output system response magnitude will grow unbounded at $r = 1$.

(c) At high frequency ratio, the amplification ratio converges to a value of one.

(d) The amplification ratio always decreases with increasing damping for all frequency ratios.

(e) At sufficiently high damping ratio, the system appears first-order-like and behaves like a high pass filter.

Significant changes are evidenced here at both high and low input frequencies when compared with the steady-state behavior of the system whose excitation magnitude is independent of frequency. In fact, the limit behavior is opposite for both systems at both low and high frequency.

Periodic Input Signal from a Base Excitation
When a system is subject to forces that are applied through its internal elements, i.e., springs and dampers by the motion of an external agent, the imposed forces are still applied by virtue of an external agent. Consider the case of an idealized model of an automobile suspension. Here, the inertial lumped mass represents the mass of a 1/4 model of an automobile comprised of a 1/4 of the chassis/frame, a single suspension strut, and one tire. The model stiffness, k, lumps together the stiffness of the suspension strut and the rubber tire while the damper primarily represents the viscous dissipation in the suspension strut.

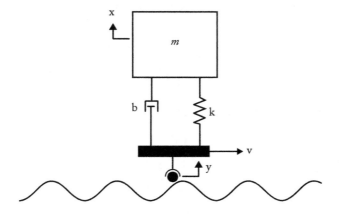

Figure 6.19: A second order system subject to excitation of its base.

The vertical motion, $y(t)$, is provided by a sinusoidal road profile with wavelength, λ, traversed by a vehicle with speed, υ:

$$y(t) = Y_0 \cos \omega t$$

and

$$\omega = \frac{2\pi \upsilon}{\lambda}.$$

A free body diagram on the inertial block renders the following governing differential equation for the horizontal motion of the mass:

$$\sum F = k\,(y(t) - x(t)) + b\,(\dot{y}(t) - \dot{x}(t)) = m\ddot{x}(t)$$

$$m\frac{d^2x(t)}{dt^2} + b\frac{dx(t)}{dt} + kx(t) = ky(t) + b\frac{dy(t)}{dt}.$$

Where the terms on the right-hand side of the equation are external forces provided by virtue of the tire motion imposed by the road profile and speed of the vehicle. Again, normalizing the governing differential equation by the system stiffness, k:

$$\frac{m}{k}\frac{d^2x(t)}{dt^2} + \frac{b}{k}\frac{dx(t)}{dt} + x(t) = y(t) + \frac{b}{k}\frac{dy(t)}{dt}$$

or, in terms of the system parameters

$$\frac{1}{\omega_N^2}\frac{d^2x(t)}{dt^2} + \frac{2\zeta}{\omega_N}\frac{dx(t)}{dt} + x(t) = y(t) + \frac{2\zeta}{\omega_N}\frac{dy(t)}{dt}.$$

Applying the Laplace operator to the normalized differential equation:

$$\left(\frac{1}{\omega_N^2}s^2 + \frac{2\zeta}{\omega_N}s + 1\right)X(s) = \left(1 + \frac{2\zeta}{\omega_N}s\right)Y(s)$$

$$G(s) = \frac{X(s)}{Y(s)} = \frac{\left(1 + \frac{2\zeta}{\omega_N}s\right)}{\left(\frac{1}{\omega_N^2}s^2 + \frac{2\zeta}{\omega_N}s + 1\right)}$$

which results in a sinusoidal transfer function

$$G(j\omega) = \frac{X_{OUT}(j\omega)}{Y_{IN}(j\omega)} = \frac{\left(1 + 2\zeta\frac{\omega}{\omega_N}j\right)}{\left(1 - \frac{\omega^2}{\omega_N^2}\right) + \frac{2\zeta}{\omega_N}\omega j} = \frac{(1 + 2\zeta rj)}{(1 - r^2) + 2\zeta rj}$$

and amplification ratio

$$\mathcal{A} = \frac{\|X_{OUT}(j\omega)\|}{\|Y_{IN}(j\omega)\|} = \frac{\sqrt{1 + (2\zeta r)^2}}{\sqrt{(1 - r^2)^2 + (2\zeta r)^2}}$$

where $r = \omega/\omega_N$, and, again, $\mathcal{A} = \mathcal{A}(r, \zeta)$.

The corresponding phase shift is given by:

$$\varphi = \angle \mathcal{N}\,(j\omega) - \angle \mathcal{D}\,(j\omega) = \tan^{-1}\,(2\zeta r) - \tan^{-1}\,\left(2\zeta r/(1 - r^2)\right).$$

So, finally

$$X_{SS}(t) = \mathcal{A}\,Y_{IN}\cos(\omega t + \varphi)$$

$$= \frac{Y_{IN}\sqrt{1 + (2\zeta r)^2}}{\sqrt{(1 - r^2)^2 + (2\zeta r)^2}}\cos\left(\omega t + \tan^{-1}\,(2\zeta r) - \tan^{-1}\,\left(2\zeta r/(1 - r^2)\right)\right).$$

The resultant Bode plot of the amplification ratio and phase shifts are shown in Figures 6.20 and 6.21, respectively.

For periodic signal input from a base excitation to a parallel mass-spring-damper system:

(a) at low frequency ratio, all of the signal input is passed onto the system with an amplification of unity and zero phase shift.

(b) At near resonant frequencies, the amplification ratio can become much larger than one. For an undamped system, the output response magnitude will grow unbounded at $r = 1$.

(c) At the peculiar frequency ratio of $r = \sqrt{2}$, the amplification ratio becomes unity irrespective of damping ratio.

(d) At high frequency ratio, the amplification ratio converges to zero.

(e) *The amplification ratio no longer decreases with increasing damping ratio for all frequency ratios!* This is true only for frequency ratios less than $r = \sqrt{2}$. For ratios higher than $r = \sqrt{2}$, increasing the amount of damping actually increases the amplification ratio. This may seem counterintuitive, but the mathematics, i.e., our "eyes with which we see physics," says it is true, and experiments verify this reality!

(f) At sufficiently high damping ratio, the system behaves like an all pass filter, i.e., the amplification ratio converges to unity for all frequency ratios.

Significant changes are evidenced here: increasing friction enhances amplification for $r > \sqrt{2}$ eventually ending up allowing all of the external excitation to be seen in the steady state at all frequencies when the friction is sufficiently high.

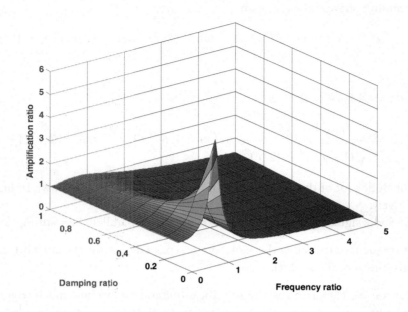

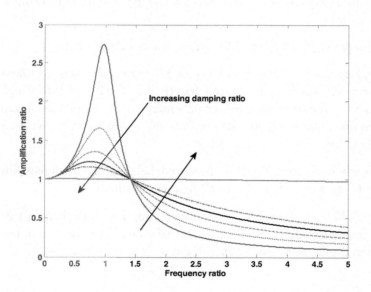

Figure 6.20: Amplification ratio for base excitation of a 2nd order system.

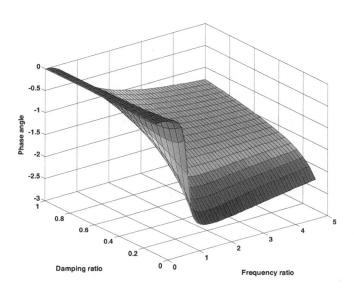

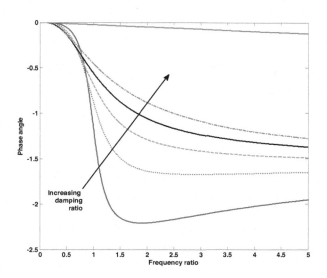

Figure 6.21: Phase shift for base excitation of a 2nd order system.

6.2.3   UNIVERSAL TRUTHS FOR 2nd ORDER SYSTEMS SUBJECT TO HARMONIC INPUT

Of Special Note

For all second order systems:

(a) Their steady-state behavior is a function of a two dimensionless parameters: the frequency ratio, $r = \omega/\omega_N$, and the damping ratio, ζ.

(b) Amplification ratios can exceed a value of unity, particularly near resonant frequencies.

(c) Phase shifts can exceed 90 degrees.

(d) Bode plots contain, at most, two inflection points allowing for peaks at intermediate frequency ratios.

  (a) Order 2 equation $\Rightarrow$ 2 inflection points

  (b) 2 inflection points $\Rightarrow$ amplification ratio and phase can increase and then decrease (or vice versa) with dimensionless frequency ratio, r.

(e) Systems can be low-pass, high-pass, mid-band pass, or all-pass filters.

6.3   REDESIGNING SYSTEMS FOR STEADY-STATE BEHAVIORS

One thing to note in second order systems is that resonance can be a particularly interesting case as amplification can be quite large. So we might want to design systems that are not capable of meandering into any troublesome regimes. Let's say, for instance, in the case of a constant force magnitude periodic input to a second order mass-spring-damper system, one wished to never see output dynamic position amplitudes greater than half the static deflection. Being interested in this limit, let's say we wish to dictate that the dynamic output be precisely half the static deflection. Recall, the amplification depicted in Figure 6.15 is a function of two parameters, the frequency ratio, $r = \omega/\omega_N$, and the damping ratio, ζ. If we limit the amplification to be precisely 1/2, then we have a unique relationship between the frequency and damping ratios shown in Figure 6.22. This figure is a cut parallel to the r–ζ plane elevated to a height of $\mathcal{A} = 1/2$.

There are now several interesting observations one can make regarding possibilities for obtaining the design condition $\mathcal{A} = 1/2$. In Figure 6.22, all $\{r, \zeta\}$ pairs to the left of the cut have amplification greater than 1/2 while all $\{r, \zeta\}$ pairs to the right have amplification less than 1/2. On the curve separating the two regions, the amplification precisely equals 1/2. If we desire

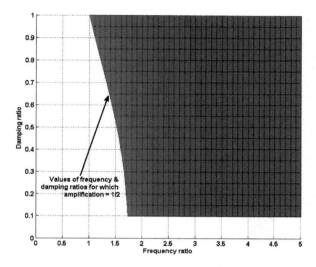

Figure 6.22: A curve $\wp(r, \zeta)$ for which $\mathcal{A} = 1/2$ for constant force magnitude periodic input to a mass-spring-damper system.

$\mathcal{A} = 1/2$, we can choose any $\{r, \zeta\}$ pair on this curve. The number of possibilities? Yes, infinity. Further, as engineers, we don't swap out parts with natural frequencies and damping ratios. We specify spring stiffnesses, damping coefficients, and inertial masses. Well, let's pick one of the infinity of solutions for a particular $\{r, \zeta\}$ pair, say $\{1, 1\}$. What specific triple $\{m, k, b\}$ corresponds to $\{r, \zeta\} = \{1, 1\}$? Well, there are, again, an infinite number of such triples that determine a single dimensionless pair. So, for every one of the infinity of $\{r, \zeta\}$ pairs, there is yet another infinity of $\{m, k, b\}$ triples! On some scale, there are ∞^2 possible solutions. Of course, one will need to consider cost, weight, availability, and other factors as constraints to fence in a reasonable solution, but there are still often a large number of potential candidates for a redesigned solution. There are a wealth of solutions at our disposal because the second order system is characterized by two independent dimensionless parameters. In first order systems, only the time constant can be changed to alter the steady-state behavior. But typically this single parameter is a product or ratio of system parameters pairs: $\{R, C\}$, $\{R, L\}$, $\{k, m\}$, $\{m, b\}$. There remain an infinite number of solutions for these pairs of system elements that will deliver the requisite time constant for a sufficient redesign of the steady-state amplitude or phase shift.

Note also that within any order system, the possibilities for redesign are dictated by the transfer function and are, therefore, dependent upon the details of the system and how it is excited by external agents. Consider that you wanted to limit the amplification ratio to a value of $1/2$ for a system exhibiting a rotating imbalance. In this case, taking the appropriate slice through the three-dimensional Bode surface results in the section shown in Figure 6.23. There are still ∞^2

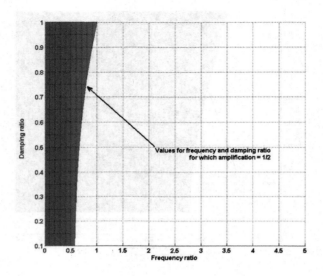

Figure 6.23: A curve $\Im(r, \zeta)$ for which $\mathcal{A} = 1/2$ for a frequency dependent magnitude periodic force input to a second order mass-spring-damper system.

potential solutions. Note that unlike the case of constant magnitude periodic force, however, now as one increases the damping ratio, the frequency ratio must increase rather than decrease in order to maintain a level amplification ratio of $1/2$. The frequency content in the magnitude of the force imbalance alters redesign scenarios in a significant way. If one increased the damping ratio along with the frequency ratio in the case where periodic force magnitude is constant, one would climb the amplification surface to values in excess of the desired design value of $1/2$. One must move, in some sense, in the opposite direction in one case than the other to achieve the desired results. Therefore, accurately modeling the system and transfer function characteristics is crucial when redesigning such dynamic systems.

6.4   ENERGY STORAGE AND DISSIPATION IN 2nd ORDER SYSTEMS SUBJECT TO HARMONIC INPUT EXCITATION

Again, system flow or effort variables solutions are calculated as primary variables. The transmitted forces and stored energies tell a part of the story not addressed by flow variables alone. For this reason, we consider the classic case of the second order mass-spring-damper subject to a constant amplitude periodic force excitation. And we will examine the energy stored by the system when the excitation frequency is low, moderate (near resonance), and high as depicted in Figure 6.24.

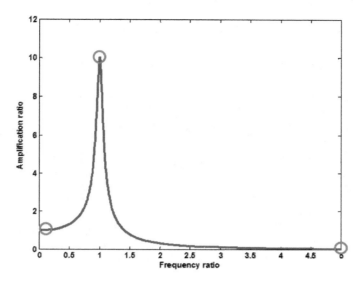

Figure 6.24: Low, resonant, and high frequency constant magnitude periodic input forces to second order system.

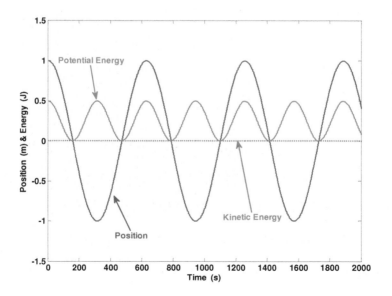

Figure 6.25: Low frequency response for position and energy in an underdamped 2nd order system subject to periodic force input of constant magnitude.

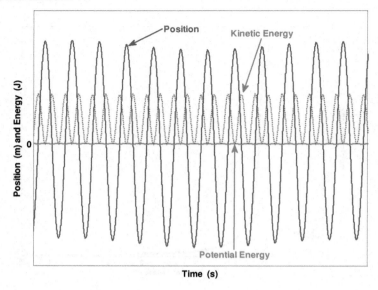

Figure 6.26: High frequency response for position and energy in an underdamped 2nd order system subject to periodic force input of constant magnitude.

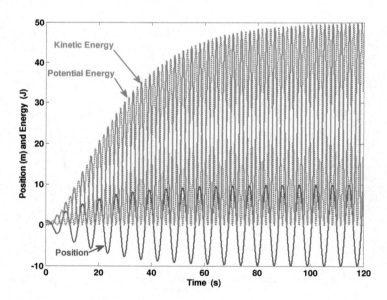

Figure 6.27: Resonant frequency response for position and energy in an underdamped 2nd order system subject to periodic force input of constant magnitude.

Because steady-state solutions are sinusoidal functions, speed is proportional to frequency. At low frequency, this minimizes the kinetic energy, leaving the lion's share of energy stored as potential energy (see Figure 6.25). Conversely, for high frequency input, the system amplitude approaches zero, leaving minimal potential energy storage. High frequency imparts high velocities and the kinetic energy is the prime storage mechanism at high frequencies (see Figure 6.26). At near resonant frequencies, both the steady-state amplitude and speed grow to large values. Here the stored energy cache takes on large values that alternate between potential and kinetic forms as shown in Figure 6.27.

6.5   CHAPTER ACTIVITIES

Problem 1 Consider the circuit pictured below, in which the bulb acts a resistor. At $t = 0$, a periodic voltage, V_0, is applied to the circuit by connecting it suddenly across a frequency modulated battery:

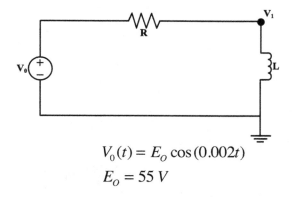

$$V_0(t) = E_o \cos(0.002t)$$
$$E_O = 55 \, V$$

The differential equation governing current response of the circuit is given by:

$$\frac{L}{R}\frac{di_L}{dt} + i_L = \frac{V_0(t)}{R} \qquad i_L(t = 0) = 0 \, \text{amps}$$

(a) Derive the amplitude ratio $\dfrac{I}{E_0/R}$.

(b) When $R = 1000 \, \Omega$, and $L = 125 \, \text{mH}$ is the electrical circuit current response fast or slow given the forcing function? Explain your answer.

(c) Redesign the series LR circuit so that the steady-state circuit current has a magnitude of 40 milliamps when driven by the periodic circuit voltage $V_0(t) = 55 \cos(0.002t) \, \text{V}$.

Problem 2 Consider the mass-less spring-damper mechanical system:

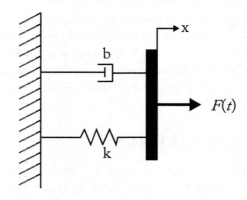

where $b = 250\,\text{Ns/m}$; $k = 125\,\text{N/m}$; $F(t) = F_0 \cos(0.25t)$; $F_0 = 100\,\text{N}$.

The differential equation governing the position of the mass-less platform is given by:

$$\frac{b}{k}\frac{dx}{dt} + x = \frac{F(t)}{k} \qquad x(t = 0) = 0\,m$$

(a) Derive the transfer function $\dfrac{X_{OUT}}{F_0/k}$

(b) Redesign the spring-damper system by changing out the spring so that the steady-state output magnitude is $0.40\,m$ when driven by the force $F(t) = 100 \cos(0.25t)\,\text{N}$.

Problem 3 Consider a bell is modeled as a cone as shown here:

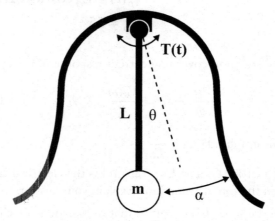

In large bells, the clapper is motor controlled at the pivot. Consider the case where the motor provides a torque given by:

$$T = 45 \cos t \,\text{Nm}$$

and there is a torsional spring with negligible damping at the pivot. Assume $m = 25$ Kg; $L = 1$ m; $\kappa = 100$ Nm/rad; $\alpha = 30°$.

(a) Assuming the mass of the rod holding the clapper is negligible, determine if the steady-state motion of the clapper will ring the bell. If yes, why? If no, why not? Assume that the shape of the bell is a cone.

(b) For what clapper mass would the steady-state motion of the clapper *just barely* reach the conical bell to ring it?

Problem 4 Consider the lumped rotational mechanical system consisting of a point mass, m, suspended at the end of a long, thin bar whose mass is lumped entirely with the point mass a distance L away from a frictionless pivot. A translational spring and dashpot are attached to the mass a distance L away from the same pivot as shown:

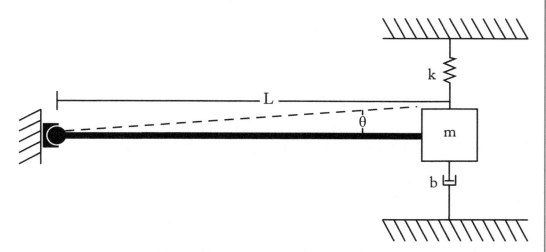

Use $m = 0.1$ kg; $k = 40$ N/m; $b = 3$ N-s/m; $L = 1$ m.

(a) Derive the differential equation governing the angular motion of the system as a function of time. Assume "small" angles to linearize the system.

(b) What are the system's natural frequency and damping ratio?

(c) Derive the system transfer function, $\dfrac{\Theta_{OUT}}{T_0/\kappa}$, if a driving torque of $T_0 = 15\cos(5t)$ Nm is applied at the pivot point, P.

(d) Identify for what non-zero input frequency, ω, the transfer function equals unity ($= 1$), i.e., the dynamic steady amplitude of angular vibration just equals the static angular deflection or "you get *out* precisely what you put *in*."

Problem 5 Consider the situation of drug absorption into a human being as mentioned in Problem 11 in Chapter 5. The human body is your system and a drug is administered by the outside world at a rate given by $f(t)$. For such a case, the differential equation governing the amount of medicine in the bloodstream, $\mathcal{M}$ is given by:

$$\frac{d\mathcal{M}}{dt} + r\mathcal{M} = f(t) \qquad t \text{ in hours}$$

where $r = 0.0833 \, \text{hr}^{-1}$.

There are two means of drug delivery: (i) by injection or (ii) a periodic dosage of so many pills per day, i.e., a periodic input. For the case of an injection of 7 mg of drugs, we have $f(t) = 0$ and $\mathcal{M}(t = 0) = \mathcal{M}_0 = 7$ mg. The pill dosage can be modeled by a periodic input:

$f(t) = 8r + 3r \cos\left(\frac{\pi}{4}t\right)$ mg/hr (t in hours), and $\mathcal{M}(t = 0) = \mathcal{M}_0 = 0$ mg.

(a) Compute the total solution for the amount of drug in the body over time for the injection and the periodic pill dosage.

(b) Compare the two solutions graphically. What amount of injection may deliver an equivalent amount of drug dosage as the pill prescription over time?

(c) In this system are you supposedly "in control" of the system variables or the outside world?

Problem 6 Consider the windshield wiper mechanism illustrated here. The mass-less blade is rigidly attached to the disk of radius R. Use $I \approx mR^2$ for the disk and wiper blade assembly for all calculations.

(a) Assuming the angular rotation of the disk remains "small," derive the differential equation governing the sweep of the wiper blade.

(b) Based on your differential equation, compute theoretical expressions for the system's natural frequency and damping ratio.

(c) Specify the damping coefficient "b" necessary if you desire the steady-state wiper blade sweep to be $\Delta\theta = \pm 45°$.

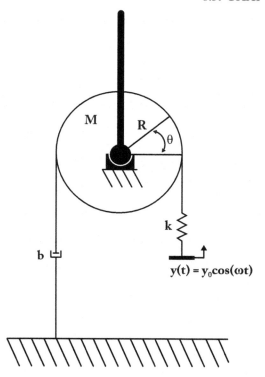

For all calculations, use:
$Y_{IN} = 0.25$ inches; $\omega = 6$ rad/s; $R = 0.5$ inches; $k = 1$ lb/ft; $m = 0.02$ slug.

Problem 7 Consider the parallel RLC electrical circuit shown below:

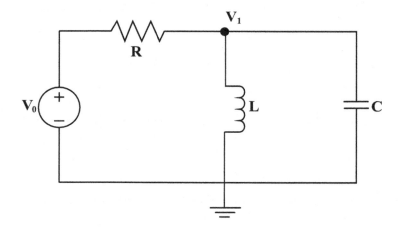

The governing differential equation for the capacitor voltage can be shown to be:

$$LC\ddot{V}_1 + \frac{L}{R}\dot{V}_1 + V_1 = \frac{L}{R}\dot{V}_0.$$

Consider that the system is excited by a frequency modulated input voltage, $V_0(t) = E_0 \cos \omega t$.

(a) Derive the transfer function for $V_1(s)/V_0(s)$ as a function of the relevant system parameters and the frequency of excitation, ω.

(b) Using the transfer function, derive an expression for the amplitude ratio V_1/E_0.

(c) Describe the behavior of the magnitude of the voltage across the capacitor at *low* frequency, resonance and *high* frequency?

(d) At resonance, for what damping ratio will the amplitude ratio, V_1/E_0, fall below unity?

(e) When the damping ratio $\zeta = 1/2$, a plot of amplification ratio vs. frequency ratio is shown here:

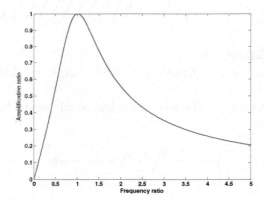

For this level of damping, determine for what input frequency ranges the output signal voltage drops below 20% of the input battery voltage if the system has a natural frequency of 2000 rad/s. What filtering characteristics would you say this system exhibits? Describe whether a first order system could exhibit such characteristics. If so, why? If not, why not?

Problem 8 Consider a downhill skier skiing down a series of moguls wherein the angle of inclination of the skier varies harmonically such that

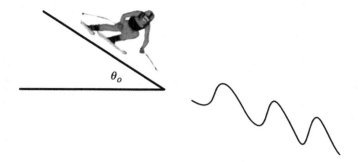

$$\theta_0(t) = 0.35 \cos 2\pi t \text{ radians.}$$

The ODE governing the skier's velocity was given by:

$$m\dot{v} + bv = mg \sin \theta_0.$$

Assuming the angle of inclination remains small, and invoking the small angle approximation:

$$m\dot{v} + bv = F(t) = mg \sin \theta_0(t) \approx mg\theta_0(t) = 0.35mg \cos 2\pi t.$$

(a) Derive the transfer function $\dfrac{V(s)}{(F(s)/b)}$.

(b) What is the steady-state magnitude of the skier's downhill velocity?

(c) With $m = 80\,\text{kg}$, and $b = 16\,\text{Ns/m}$, classify the *mogul gravity loading* as low ($\tau\omega \ll 1$), intermediate ($\tau\omega \approx 1$), or high frequency ($\tau\omega \gg 1$).

(d) Show that for a "very heavy" skier, their steady-state velocity magnitude would be:

$$V_{SS} = \frac{0.35g}{2\pi}$$

i.e., that the steady-state velocity magnitude of the skier is independent of the skier's mass. Is the skier described in part (c) "heavy" in steady state, dynamically speaking?

Problem 9 Consider the translational series mass-spring-damper mechanical system shown below forced by excitation of the damper $y(t) = Y_{IN} \cos(\omega t)$:

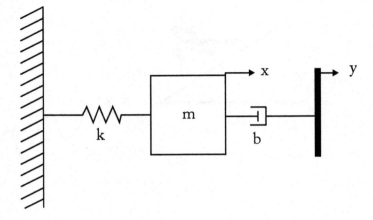

(a) Show that the governing differential equation is given by:

$$m\ddot{x} + b\dot{x} + kx = b\dot{y}$$

(b) Using the transfer function, derive an expression for the amplitude ratio $\dfrac{X_{OUT}}{Y_{IN}}$ in terms of the damping ratio and the dimensionless frequency ratio. Describe, in words, the behavior of the amplification ratio at low ($r \ll 1$), intermediate ($r \approx 1$), and high ($r \gg 1$) frequencies.

(c) Is this system analogous to any of the electrical circuits you have experienced thus far? If so, describe the analogous system elements for each.

(d) If the damper is pumped at a frequency $\sqrt{2}$ times the system's resonant frequency, i.e., $y(t) = Y_{IN} \cos \sqrt{2}\omega_N t = 4 \cos \sqrt{2}\omega_N t$ and the damping ratio for the system is $\xi = 1/2$, determine the output, steady-state motion of the mass, m, as a function of time.

(e) At resonance, for what damping ratio will the amplitude ratio, $\dfrac{X_{OUT}}{Y_{IN}}$, fall below unity?

Problem 10 Consider the regenerative braking lumped model illustrated here. Pumping the brake pedal effectively acts as a base excitation on a damper linked to the brake disk of diameter, D, and mass, $m = 0.25$ slugs.

$k = 4\,\text{lb/ft}$; $b = 20\,\text{lb-s/ft}$; $D = 1\,\text{ft}$; $J = \frac{1}{2}mR^2$; $y(t) = 0.5 \cos(20t)$ ft.

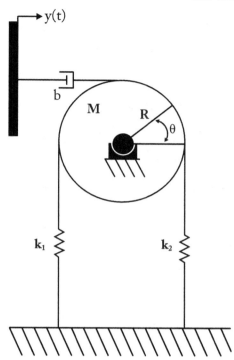

Show that, when the angular motion of the disk remains small, the governing ODE for the angular motion of the disk is given by:

$$J\ddot{\theta} + bR^2\dot{\theta} + 2kR^2\theta = Rb\dot{y}$$

where $J = \frac{1}{2}mR^2$.

(a) Derive the transfer function for $\dfrac{\Theta_{OUT}}{Y_{IN}/D}$.

(b) Using $k = 4\,\text{lb/ft}$; $b = 20\,\text{lb-s/ft}$; $D = 1\,\text{ft}$; $m = 0.25\,\text{slugs}$; $y(t) = 0.5\cos(20t)\,\text{ft}$. what is the steady-state angular motion amplitude, Θ_{OUT}?

(c) Redesign the dashpot to reduce the steady-state amplitude to 0.25 radians.

CHAPTER 7

The Fluid and Thermal Casts

Finally, we introduce two last casts of characters telling the story of effort and flow in fluid and thermal systems.

7.1   FLUID SYSTEMS

The flow of fluids fascinates everybody. We watch streams, waterfalls, whirlpools, and we are fascinated by this substance which seems almost alive relative to solids.

Richard P. Feynman
The Feynman Lectures on Physics

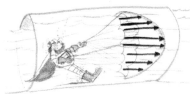

The hydraulic analogy compares electric current flowing through circuits to water flowing through pipes. When a pipe is filled with hair, it takes a larger pressure to achieve the same flow of water. Pushing electric current through a large resistance is like pushing water through a pipe clogged with hair: It requires a larger push or voltage drop to drive the same flow or electric current.

Wikipedia

Have you ever wondered why water is stored in high towers or standpipes? By virtue of their height, towers storing fluid produce hydrostatic pressure sufficient to drive the fluid out into distribution systems such as pipes for homes and businesses. Fluid flows out of the tank under the gravitational force of its own weight. The fluid effort across a volume of contained fluid pushes the fluid which then responds by flowing. Again, your intuition helps in this telling of the story.

7.1.1   FLUID EFFORT AND FLOW VARIABLES

What pushes fluid is a pressure differential across, say, a length of pipe. This drives a volume flow rate of fluid, Q, through the pipe. At their essence, fluid systems are special cases of mechanical systems in general. As with mechanical systems, when the fluid is incompressible, this volume flow rate directly implies a mass flow rate.

$$\dot{m} = \frac{dm}{dt} = \rho Q.$$

Table 7.1: Effort, flow, and conserved quantities for fluid systems

Conserved Quantity		Units	Symbol
Fluid mass		kg	m
Variable		Units	
Effort	Pressure	N/m^2 ; lb/ft^2	p
Flow	Volume flow rate	m^3/s ; ft^3/s	$Q = dV/dt$

7.1.2 STORAGE ELEMENTS

The fluid cast is capable of storing energy in both potential and kinetic forms. The fluid system is nearly always a circuit of containment vessels that deliver fluid from one location to another.

Potential Energy Storage Character

Potential energy storage in fluid systems takes place when the fluid stores a large effort or pressure differential in a fluid circuit. The fluid cast member who plays the role of Captain Potential Energy is a storage tank. By virtue of a height or head of fluid, a large static pressure differential is built up due to gravitational loading. Let's now imagine: what factors will determine the amount of potential energy that a tank can store? It seems intuitive that the volume of fluid may matter. But fluid volume in, say, a cylindrical tank is a product of its area and height. It is a result from fluid statics that the pressure at the bottom of a column of fluid is determined solely by the height of fluid in the column. Often pressure is measured in pressure head or the height of liquid of a given density that produces a given pressure. Pressure and height are, in this sense, both equally interchangeable effort variables. Since pressure, and not force, drives the fluid mass, what properties of a storage tank make for a fluid system having the capacity to drive flow of fluid mass?

First, let's follow the mathematical relation for storage of potential energy of water kept in a tank of cross-sectional area, A. As we have already mentioned, the pressure at the bottom of the tank will be related to the height of water in the tank. So pressure and height are interchangeable effort variables. For the moment, let's focus on pressure itself. Since you have not had a course in fluid mechanics yet, let's practice letting the mathematics and our analogy guide us. The mathematical expression of the storage by virtue of effort is

$$FLOW = C_{FLUID} \frac{d(EFFORT)}{dt}.$$

In this way, we have

$$C_{FLUID} = \frac{Q}{dp/dt} = \frac{dV/dt}{dp/dt} = \frac{dV}{dp}.$$

Here, the fluid capacitance, C_{FLUID}, is a rate of change of fluid volume corresponding to a rate of change in applied pressure. Fluid dynamicists refer to this quantity as the fluid compliance. Then

Figure 7.1: The fluid potential energy storage character is played by the storage tank. It stores energy in potential form in accordance with increased height of mass in the tank and storage of a pressure differential across the height of the tank.

the analogy with mechanical systems comes full circle because in mechanical systems, the inverse of a substance's stiffness is its compliance

$$C_{MECH} = k^{-1}.$$

So analogously for fluid systems

$$dp = d\left(\frac{mg}{A_{TANK}}\right) = \frac{g}{A_{TANK}}dm = \frac{g}{A_{TANK}}d(\rho V) = \frac{\rho g}{A_{TANK}}dV$$

$$C_{FLUID} \equiv \frac{dV}{dp} = \frac{A_{TANK}}{\rho g}$$

where the fluid capacitance is measured in

$$C_{FLUID} \doteq \mathrm{m^4 s^2/kg}.$$

Kinetic Energy Storage Character

When considering energy storage via flow, fluid systems are directly analogous with translational mechanical flow. The fluid cast member who plays the role of Captain Kinetic Energy is that device that stores energy by virtue of its volume flow rate. Consider the case of a fluid that is incompressible. The volume flow rate in a cylindrical pipe is determined directly by the fluid velocity along the pipe. Kinetic energy is stored by virtue of fluid velocity that is, in the strictest sense of our analogy, stored by a measure of the fluid inertia. In fluid systems, this is often referred to as the fluid inertance.

Again, without a physical intuition or feel for inertance, let's allow the analogy to guide us mathematically. This may seem abstract, at the moment, but the analogous behavior, in the

Figure 7.2: The fluid kinetic energy storage character is played by the system's inertia. Fluid inertance is embodied in a fluid system's mass.

end, will hopefully bolster our physical feel once we undertake a course in fluid mechanics and dynamics. The mathematical expression of the storage by virtue of flow is

$$EFFORT = \mathcal{L}\frac{d(FLOW)}{dt}$$
$$p = \mathcal{L}\frac{dQ}{dt}.$$

Understanding that fluids are a special case of mechanical systems

$$F = m\frac{dv}{dt}.$$

Using $F = pA$, $m = \rho A\ell$ and $Q = Av$

$$pA = \rho A\ell\frac{dv}{dt} = \rho\ell\frac{dQ}{dt}$$

or for fluid, say, flowing in a pipe of length, ℓ_{PIPE}

$$p = \frac{\rho\ell}{A}\frac{dQ}{dt} = \mathcal{L}\frac{dQ}{dt}$$
$$\Rightarrow \mathcal{L}_{FLUID}^{PIPE} = \rho_{FLUID}\,\ell_{PIPE}/A_{PIPE}$$

where the fluid inertance is measured in

$$\mathcal{L}_{FLUID} \doteq \frac{\text{kg}}{\text{m}^4}.$$

Many fluid systems are designed for steady flow purposes, e.g., hoses, faucets, pipelines. Transients occur when such systems are turned on and shut off, but for most of the operating time, the flow is steady and $\dot{Q} \propto dv/dt \approx 0$. In these instances, inertia plays a negligible role in the energy storage and the inertance is then neglected.

7.1.3 DISSIPATIVE ELEMENTS

Energy dissipation in fluid systems results from any element in the fluid circuit that impedes fluid flow rate. The role of the Evil Dr. Friction in the fluid flow script is played by the physical presence of friction acting against the flow of fluid. Two salient examples are pipe friction and losses exhibited in flow of fluid through valves or constrictions.

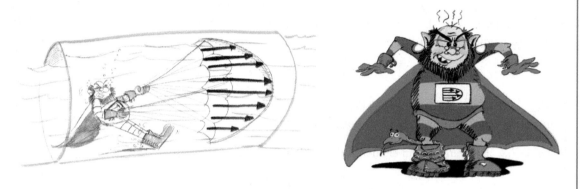

Figure 7.3: The friction force is modeled by the net viscous force that is proportional to a pressure difference in the fluid circuit.

The governing mathematical expression of the dissipation is algebraic and often bears someone's name! Let's consider an incompressible, viscous fluid undergoing slow, laminar flow in a pipe. For such conditions, the Hagen-Poiseuille law relates volume flow rate, Q, of the fluid to the pressure difference applied across the section of pipe driving the flow

$$p = RQ \Rightarrow R = p/Q.$$

And the Hagen-Poiseuille flow law is given by

$$Q = p\pi R D^4/128\mu\ell$$

where μ is the viscosity of the fluid, and D and ℓ are the diameter and length of the pipe respectively. The viscosity is a fluid property that quantifies a fluid's material resistance to flow. It is measured in poises:

$$1\,poise \doteq 0.1\,\text{Ns/m}^2.$$

Using our analogy for resistance:

$$EFFORT = R * FLOW$$
$$p = RQ$$

and for Hangen-Poiseuille flow

$$p = \frac{128\mu\ell}{\pi D^4}Q$$
$$\Rightarrow R_{FLUID}^{PIPE} = \frac{128\mu\ell}{\pi D^4}.$$

The resistance to flow will increase linearly with the pipe length and fluid viscosity. The resistance will also decrease as the pipe radius is increased, but this dependence is to the fourth power! The fluid resistance is measured in

$$R_{FLUID} \doteq \frac{Ns}{m^2} * m/m^4 \doteq \frac{kg}{m^4 s}.$$

Flow resistances from higher velocity flows must account for turbulence. These resistances are almost always nonlinear and will not be considered explicitly here.

Table 7.2: Relevant system element relations for fluid systems

Field	Effort Variable	Flow Variable
Fluid	Pressure	Mass flow rate
Relation	Form	Analogy
Dissipative Material Property Law	Effort = Resistance x Flow Linear $(p_1 - p_2) = RQ$	Resistance = Laminar Pipe Flow Linear Resistance = $\dfrac{128\mu L}{\pi D^4}$
Energy Storage in Effort Variable	Flow = Capacitance x d(Effort)/dt $Q = \dfrac{A}{\rho g}\dfrac{dp}{dt}$	Fluid Capacitance = Compliance $C_{FLUID} = \dfrac{A}{\rho g}$
Energy Storage in Flow Variable	Effort = Inductance x d(Flow)/dt $p = \mathcal{L}\dfrac{dQ}{dt}$	Fluid Inductance = Inertance $\mathcal{L}_{FLUID}^{PIPE} = \dfrac{\rho_{FLUID}\ell_{PIPE}}{A_{PIPE}} =$

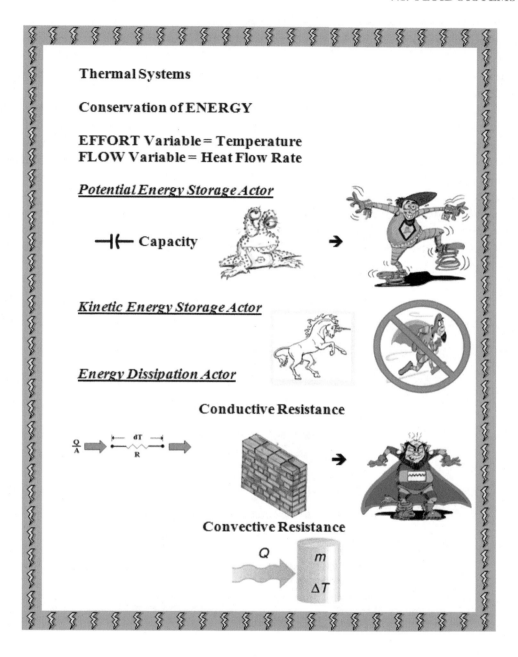

Figure 7.4: The fluid system cast of characters.

7.1.4 SINGLE STORAGE ELEMENT SCRIPTS

An idealized case often studied is that of the storage tank draining out of an aperture cut below the fluid surface or into an exterior pipe. Here, we might be asking how much time it takes to fill or drain the tank. Or we might be interested in calculating the height of fluid in the tank under steady flow conditions.

The system is comprised of the standing tank acting as the fluid capacitor, and the draining pipe which is the fluid resistor. You may ask why the tank's resistance is not considered. It is, after all, a sort of "short" pipe with a rather large diameter. But consider the ratio of the tank's effective length to its diameter to the fourth power. When this value is negligible compared to that of the drainpipe, then the resistance of the pipe dominates over that of the tank and it may be reasonable to neglect the flow resistance of the tank.

We perform a force balance on a representative control volume of fluid in the pipe. Father Force, pictured on the ladder in Figure 7.5, provides a supply of water from the outside world. Let's presume he turns on an input tap that provides a fluid volume flow rate of Q_{IN}. The pressure

Figure 7.5: The classic problem of the draining tank.

difference across the pipe created by the weight of fluid in the tank drives the outgoing flow in the drainpipe. The pressure at the free surface in the tank and the outflow of the pipe is atmospheric. If we use this value as a reference effort value, or alternatively use the so-called gauge pressure, we can set these reference values of pressure to zero. Then the operative pressure difference across the pipe is illustrated in Figure 7.6.

Figure 7.6: Mass flow rate over a control volume of fluid in the draining pipe.

From the effort flow analogy

$$Q_{IN} - Q_{OUT} = C_{FLUID}\frac{dp}{dt}.$$

The input volume flow rate is externally provided by "the outside world," aka Father Force. The output volume flow rate depends on the resistance of the pipe while the capacity to maintain a driving pressure difference is determined by the characteristics of the storage tank. Using the corresponding system element equations corresponding to Dr. Friction and Captain Potential

Energy, respectively,

$$Q_{IN} - p/R_{FLUID}^{PIPE} = C_{FLUID} \frac{d(p)}{dt}$$

$$R_{FLUID}^{PIPE} C_{FLUID}^{TANK} \frac{dp}{dt} + p = R_{FLUID}^{PIPE} Q_{IN}.$$

This is a differential equation for the pressure at the bottom of the tank or entry to the pipe, the system effort variable. This is also linearly related to the height of fluid in the tank, often called the *pressure head*.

$$p = \rho A h g / A = \rho g h.$$

Performing a change of variable from pressure to pressure head

$$R_{FLUID}^{PIPE} C_{FLUID}^{TANK} \frac{d(\rho g h)}{dt} + \rho g h = R_{FLUID}^{PIPE} Q_{IN}$$

$$R_{FLUID}^{PIPE} C_{FLUID}^{TANK} \frac{dh}{dt} + h = R_{FLUID}^{PIPE} Q_{IN} / \rho g.$$

When there is no source from the outside world, the equation will be homogeneous. The solution of the homogeneous equation is the sole transient and the steady state is an empty tank with zero height of fluid and zero gauge pressure. When there is an external flow source, the steady-state height of fluid in the tank will coincide with the condition that

$$dh/dt = 0 \Rightarrow h_{SS} = R_{FLUID}^{PIPE} Q_{IN} / \rho g.$$

The time constant is given by the classical RC expression using the hydraulic analogy to electrical systems

$$R_{FLUID}^{PIPE} C_{FLUID}^{TANK} = \frac{128 \mu \ell_{PIPE} A^{TANK}}{\rho g \pi D_{PIPE}^4} = \tau.$$

Note that τ turns out to have units of time, as we expect from the analogy:

$$\tau \equiv R_{FLUID}^{PIPE} C_{FLUID}^{TANK} \doteq \left(\frac{kg}{m^4 s} \right) m^4 s^2 / kg \doteq s.$$

Recall that the governing equations for electrical systems typically appear in terms of effort and/or flow while mechanical systems are most often in terms of flow. Steady incompressible fluid systems are most often written in terms of effort, either the fluid pressure or pressure head.

7.1.5 MULTIPLE STORAGE ELEMENT SCRIPTS

A multiple storage script must involve fluid kinetic energy as well as fluid potential energy. An illustrative case is that of the U-tube manometer. Fluid in static equilibrium in a vertical U-tube will contain as much fluid mass or climb as high in the left tube as the right tube as shown in

Figure 7.7. If an external pressure were applied to the free surface in the left tube, a relative fluid height would develop as the fluid originally in the left tube is displaced into the right tube. If the pressure were then released, this displaced fluid would then be driven by a net gravitational loading until it moved back into the left tube. This motion would resemble that of a pendulum released from a given initial angle.

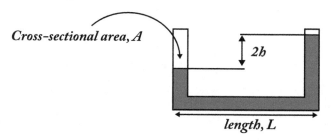

Cross-sectional area, A

2h

length, L

Figure 7.7: A classical U-tube manometer fluid pendulum.

If the tube friction is not sufficient to prevent it, the fluid will overshoot the original equilibrium position by virtue of the kinetic energy of flow. Then it will climb up the left tube and "swing" back and forth as a fluid pendulum. Friction between the fluid and the tube walls will provide damping and the transfer of potential energy to kinetic energy and back will be accompanied by losses that cause the fluid pendulum swing to eventually cease (see Figure 7.8).

Writing a momentum balance on a representative fluid control volume, one can show that the differential equation governing the relative height of fluid in the manometer is given by:

$$\rho A L \frac{d^2 h}{dt^2} + R_{FLUID} A^2 \frac{dh}{dt} + 2\rho g A h = AP(t)$$

where $P(t)$ is an externally imposed gauge pressure at one fluid surface. If we scale the entire equation to normalize the effort variable term of pressure head, one can show that (see Chapter Activities Problem 2):

$$\mathcal{L}_{FLUID} C_{FLUID} \ddot{h} + R_{FLUID} C_{FLUID} \dot{h} + h = H_O(t)$$

where the system element equation for the capacitance of a U-tube manometer is

$$C_{FLUID} = A/2\rho g$$

and the pressure head forcing function is

$$H_0(t) = P(t)/2\rho g.$$

You should take note that the coefficient of the head, h, is already unity. As such, we have:

$$L_{FLUID} C_{FLUID} \doteq \frac{kg}{m^4} \frac{m^4 s^2}{kg} \doteq s^2$$

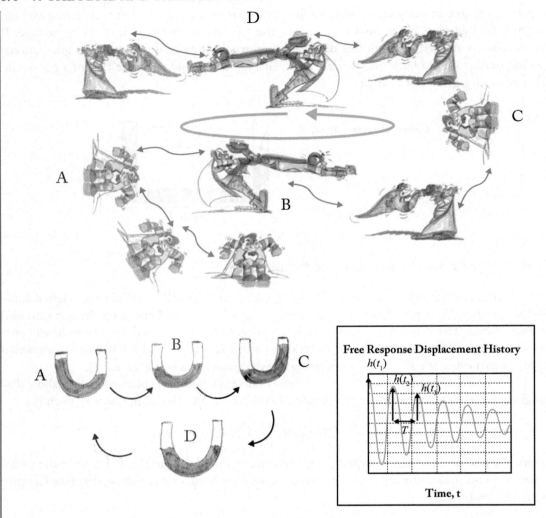

Figure 7.8: The energy catch with losses in a U-tube manometer fluid pendulum.

which exhibits units of $1/\omega_N^2$ and

$$R_{FLUID}C_{FLUID} \doteq \text{s}$$

which exhibits units of $2\zeta/\omega_N$.

In this script, the externally applied pressure drives the fluid mass which is initially opposed by a gravitational spring and tube friction. As the kinetic energy imparted to the mass by the pressure is reduced, an equivalent amount of potential energy is stored in the spring or height in the remaining tube. The fluid system energy is simply transferred from kinetic to potential and back with dissipation provided by the tube walls.

Once again, Captain Potential Energy and Captain Kinetic Energy "have a catch" with a ball of energy while the Evil Dr. Friction takes a bite at each pass.

Figure 7.9: A dynamic second order system energy exchange with dissipation.

7.2   THERMAL SYSTEMS

There is apparently no thermal element which displays an energy storage mechanism which is complementary to the flow store.

Paul E. Wellstead
Introduction to Physical System Modeling

But as sure as you're born … you're never gonna see no unicorn.

Shel Silverstein
"The Unicorn"

Finally, we introduce our last cast of characters telling the thermal story of effort and flow. The effort-flow analogy holds only in part for thermal systems because Captain Kinetic Energy does not exist! There is no storage nor balance of a momentum-like quantity in any thermal system. Thermal kinetic energy is a unicorn. In other words, you won't find one!

7.2.1   THERMAL EFFORT AND FLOW VARIABLES

In thermal systems, your intuition will again serve you well. You already know that a temperature difference across an element will cause heat to flow *from hot to cold*. Therefore, temperature plays the role of effort while heat flow rate is the flow variable.

Table 7.3: Effort, flow, and conserved quantities for thermal systems

Conserved Quantity		Units	Symbol
Heat energy		Joules	J
Variable		Units	
Effort	Temperature	^{o}C ; ^{o}F	T
Flow	Heat flow rate	Watt = J/s; BTU/hr	q

7.2.2   STORAGE ELEMENTS

The single most interesting characteristic of thermal systems is, arguably, that they can store only one type of energy, namely potential. This is the main and a crucial difference between thermal and all other systems. As such, thermal systems can only ever be governed by first order differential equations in time. Let's examine the potential energy character in detail.

Potential Energy Storage Character

Thermal capacitance is defined as the capacity to store an effort differential across an element. Here that translates into a temperature difference. The energy stored per unit temperature difference is a measure of the capacitive strength. The thermal cast member who plays the role of Captain Potential Energy is the mass which provides material-specific heat capacity.

The mathematical expression of the storage by virtue of effort is

$$FLOW = C_{THERM}\frac{d(EFFORT)}{dt}$$

which can then be written for thermal systems

$$q = mc_P\frac{d(\Delta T)}{dt}$$

$$q = C_{THERM}\frac{d(T - T_{REF})}{dt} = mc_P\frac{dT}{dt}$$

$$\Rightarrow$$

$$C_{THERM} \equiv mc_P.$$

Here, our thermal capacitor represents the potential for a thermal system to store thermal heat energy by virtue of a temperature difference contained in the element. A material's ability to store

Figure 7.10: The thermal potential energy storage character is played by the heat capacity carried by the system mass. It embodies the thermal capacitance of the system.

heat energy for every degree of rise in its temperature is referred to as its *heat capacity*, c_P.

$$q = mc_P \frac{d(\Delta T)}{dt}$$

$$q = C_{THERM} \frac{d(T - T_{REF})}{dt} = mc_P \frac{dT}{dt}$$

where the quantity $mc_P (T - T_{REF})$ is referred to as the internal energy of the system. The thermal capacitance is the total thermal heat capacity

$$C_{THERM} \equiv mc_P.$$

The heat capacity is an extensive quantity and is proportional to the system's thermal mass. It is a measure of how much energy can be stored in a mass before its temperature will increase a single degree:

$$mc_P \doteq \text{kg} \frac{\text{J}}{\text{kg} - {}^\circ\text{C}}.$$

Kinetic Energy Storage Character

In Shel Silverstein's words, "you're never gonna see no unicorn." Thus, is the thermal kinetic energy storage character. Kinetic energy elements that store energy by virtue of the flow variable simply do not exist. Ergo, Paul Wellstead's notion that "apparently, there are none." Captian Kinetic Energy is AWOL! This has tremendous implications for thermal system dynamics. Namely, because both Captain Potential Energy *and* Captain Kinetic Energy must be present and accounted for in order to have a second order system, all thermal systems are necessarily governed by first order equations in time.

Figure 7.11: The thermal kinetic energy storage character does not exist!!

7.2.3   DISSIPATIVE ELEMENTS

Dissipation in thermal systems is provided by physical agents that impede heat flow. Heat flow is impeded differently in solid, fluid, and a vacuum. Heat flows through a solid by means of conduction, through fluids by means of convection, and through a vacuum by radiation. Radiative heat flow is highly nonlinear and will not be addressed here.

Conductive Resistance to Heat Flow

Heat flows through solids by conduction, a process in which heat thermally agitates the solid atoms in their lattice. The solid lattice impedes the flow of heat. A temperature difference must be imposed across a solid to drive heat flow through it. By virtue of their lattice structure, solids that are conducting heat provide a thermal resistance to heat flow.

Here, Fourier's law of heat conduction provides a relationship between a temperature difference across a solid of constant thickness and the resulting heat flow rate by virtue of a material property known as the thermal conductivity, k. Fourier said that the heat flux through a solid is proportional to the local temperature gradient through the thermal conductivity

$$q_{COND} = -kA\frac{dT}{dx}.$$

Consider once more that in deriving a differential equation for heat flow, we balance heat into and out of a small representative control volume in the system. If the control volume is sufficiently small, any temperature distribution will "look linear" and we can model the temperature gradient as a finite difference

$$q_{COND} = -kA\frac{dT}{dx} \approx -kA\frac{\Delta T}{\Delta x} = -kA\frac{T_2 - T_1}{x_2 - x_1} = kA\frac{T_1 - T_2}{L}$$

where L is some representative distance across which conduction is taking place through "a window" of cross-sectional area, A, and at whose ends the temperatures are T_1 and T_2, respectively.

Figure 7.12: Solids provide thermal resistance to heat flow by the energy lost through thermal agitation of strongly bonded solid lattice networks.

We are now able to relate the temperature difference necessary to drive heat flow through a resistive element

$$T_1 - T_2 = \frac{L}{kA} q_{COND} = R_{THERM}^{COND} q_{COND}$$

where we can now apply the system analogy

$$EFFORT = R_{THERM}^{COND} * FLOW$$
$$R_{THERM}^{COND} \equiv L/kA$$

where the units of thermal resistance are given by

$$L/kA \doteq m/\frac{W}{m - °C} m^2 \doteq \frac{°C}{W}$$

and heat flow rate is measured in watts

$$W \equiv Watt \doteq \frac{J}{s}.$$

Convective Resistance to Heat Flow

Alternatively, heat flow is impeded in a different manner when being transferred through a fluid. Heat flows through a fluid medium by a process known as convection, and the fluid provides a thermal resistance as heat convects through the fluid under the influence of an imposed temperature difference.

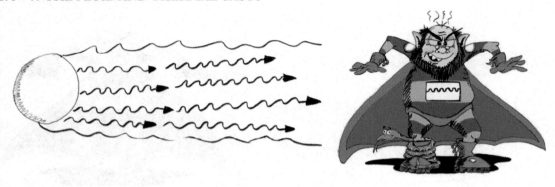

Figure 7.13: Fluid media provide thermal resistance to heat flow by the energy lost through thermal agitation of loosely bound fluid molecules.

Convective heat flow is governed by Newton's Law of Cooling whereby a solid at temperature, T, surrounded by a large reservoir of fluid at temperature, T_∞, will result in heat transferred through the fluid given by

$$q_{CONV} = hA\,(T - T_\infty) = hA\Delta T$$

where h is referred to as the heat transfer or film coefficient and A is the area through which the heat is flowing. Inverting this relationship, the resulting temperature difference between the solid surface and the fluid becomes

$$\Delta T = \frac{1}{hA}q_{CONV} = R^{CONV}_{THERM}q_{CONV}$$

and invoking the effort-flow analogy

$$EFFORT = R^{CONV}_{THERM} * FLOW$$
$$R^{CONV}_{THERM} \equiv 1/hA$$

where $R^{CONV}_{THERM} \equiv 1/hA \doteq 1/\dfrac{W}{m^2{}^\circ C}m^2 \doteq \dfrac{^\circ C}{W}$.

A list of thermal system element equations is given in Table 7.4. A summary of the thermal cast and the roles they play is given in Figure 7.14.

7.2.4   SINGLE STORAGE ELEMENT SCRIPTS

An idealized case often studied is that of conduction through a solid, insulated wall. The solid is characterized by a capacity to retain heat measured by its temperature. Father Force is now temperature. The heat capacity of the wall allows it to store thermal energy by virtue of its temperature. This is referred to as the solid wall's internal energy. The heat capacity of the wall is

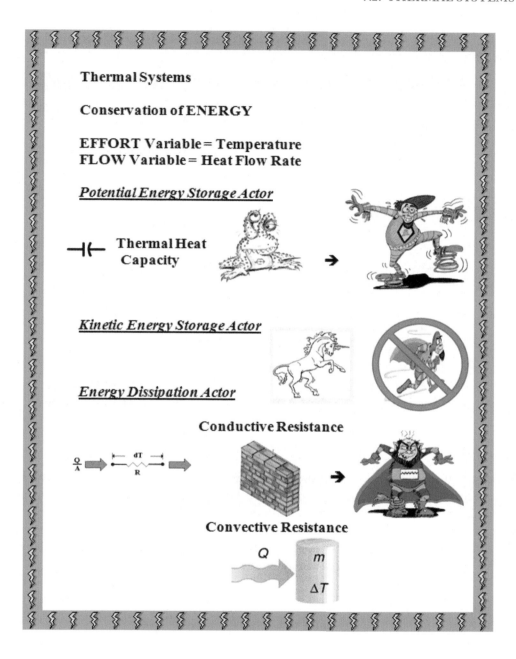

Figure 7.14: The thermal system cast of characters.

Table 7.4: Relevant system element relations for thermal systems

Field	Effort Variable	Flow Variable
Thermal	**Temperature**	**Heat flow rate**
Relation	*Form*	*Analogy*
Dissipative Material Property Law	**Effort = Resistance x Flow** $$(T_1 - T_2) = R\,q$$	**Convective** $$\text{Resistance} = \frac{1}{hA}$$ **Conductive** $$\text{Resistance} = \frac{L}{kA}$$
Energy Storage in Effort Variable	**Flow = Capacitance x d(Effort)/dt** $$q = mc_P\frac{dT}{dt}$$	**Capacitance = Thermal Heat Capacity** $$C_{THERM} = mc_P$$
Energy Storage in Flow Variable	**Effort = Inductance x d(Flow)/dt** **Not Applicable**	**Inductance =** **There is no thermal equivalent or analog for inductance**

likened to a thermal spring being pushed by Father Force as shown in Figure 7.15. This illustrates the capacity of the wall to remain at an elevated temperature and store thermal energy in a form measurable by its effort variable. The solidly bonded molecules of the insulating layer provide resistance to heat flowing through them to the outside, $T_{OUT} < T$.

In order to balance heat flow rate through the insulation, we perform a thermal heat energy balance on a representative control volume in the insulation.

$$0 - q_{OUT} = C_{THERM}\frac{dT}{dt} = mc_P\frac{d\,(T)}{dt}$$

where T is the temperature of the wall. If the dominant temperature difference is that between the wall and the temperature outside of the insulating layer, T_{OUT}, then we can represent the heat

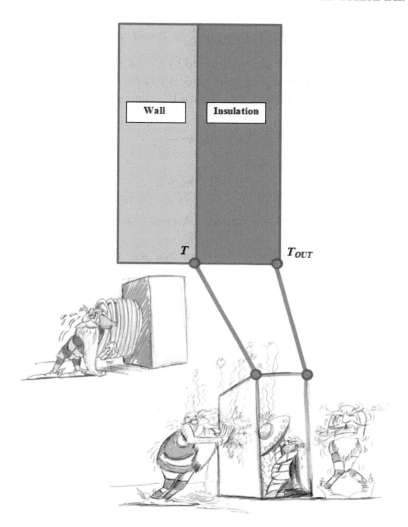

Figure 7.15: Heat flow through a control volume across a solid wall.

flowing out through the insulation as

$$-kA\Delta T/L = -(T - T_{OUT})/R_{CONDUCTION}^{INSULATION} = mc_P \frac{d(T)}{dt}$$

$$\Rightarrow R_{THERM}^{COND} C_{THERM} \frac{dT}{dt} + T = T_{OUT}$$

resulting in

$$R_{THERM}^{COND} C_{THERM} \frac{dT}{dt} + T = T_{OUT}.$$

The excitation from the outside world is provided by the external temperature. The solution of this equation is a temperature changing monotonically from $T(0)$ to T_{OUT} in roughly four time constants. The time constant is given by the classical RC expression using the analogy to electrical systems

$$R_{THERM}^{COND} C_{THERM} = \frac{m c_P L}{k A} = \tau.$$

Note that the units of the time constant are:

$$R_{THERM}^{COND} C_{THERM} \equiv \tau \equiv \frac{m c_P L}{k A} \doteq \frac{J}{^\circ C} \frac{^\circ C}{W} \doteq s$$

or units of time. The analogy delivers a parameter known to characterize all first order systems in time as we've described them.

Alternatively, a simple illustration of convective heat transfer occurs during quenching: when a hot, small solid object is transferred to a large cooling bath (Fig. 7.16). In order to balance

Figure 7.16: Heat flow through a control volume contained in a fluid surrounding an object from which heat is being transferred.

heat flow rate in the fluid surrounding the quenched sphere, we perform a heat energy balance on a representative control volume in the fluid reservoir.

$$q_{IN} - q_{OUT} = q_{STORED}$$

$$0 - hA(T - T_\infty) = m c_P \frac{dT}{dt} = C_{THERM} \frac{dT}{dt}.$$

Invoking the effort-flow analogy

$$0 - \frac{1}{R^{CONV}_{THERM}}(T - T_\infty) = mc_P \frac{dT}{dt} = C_{THERM} \frac{dT}{dt}.$$

Rearranging

$$R^{CONV}_{THERM} C_{THERM} \frac{dT}{dt} + T = T_\infty.$$

The excitation from the outside world is provided by the quench tank fluid reservoir temperature. The solution of this equation is a temperature changing monotonically from $T(0)$ to T_∞ in roughly four time constants. The time constant is given by the classical RC expression using the analogy to electrical systems

$$R^{CONV}_{THERM} C_{THERM} = \frac{mc_P}{hA} = \tau.$$

7.3   CHAPTER ACTIVITIES

Problem 1 A U-tube manometer is a relatively simple device used to measure pressure. When the fluid level is displaced as shown above and released, the following oscillation in relative fluid height is observed:

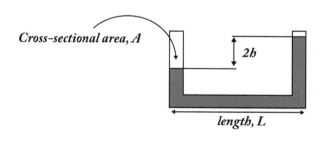
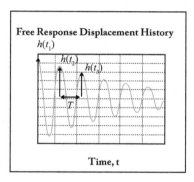

When a periodic pressure is applied at one end, a force and mass balance on the system gives the following governing differential equation for the fluid height, $h(t)$:

$$\rho A L \frac{d^2 h}{dt^2} + R \rho A^2 \frac{dh}{dt} + 2\rho g A h = PA \cos 4t$$

$$h(t = 0) = 5\,\text{cm}$$

$$\frac{dh}{dt}(t = 0) = 0\,\text{cm/s}$$

where ρ is the fluid density, $A = 1\,\text{cm}^2$, and $L = 5\,\text{cm}$.

(a) Using representative analogies in Table 7.2, show that the differential equation governing the height, h, can be written as:

$$\mathcal{L}_{FLUID}C_{FLUID}\,\ddot{h} + R_{FLUID}C_{FLUID}\,\dot{h} + h = H_0(t)$$
$$H_0(t) = P(t)/2g\rho$$
$$C_{FLUID} = A/2\rho g.$$

(b) Write an algebraic expression for the fluid inertance, i.e., the fluid inertia.

(c) Calculate the natural frequency and the fluid resistance, R, in this pendulum system if it is critically damped. Assume the acceleration due to gravity is given by $g = 10\,\text{m/s}^2$.

(d) What periodic pressure magnitude, P, needs to be applied to obtain a steady-state output height of 2 cm (an amount that will just cause liquid to spill out of the U-tube).

(e) What are the characteristic times for the system in part (d)?

(f) If the tube resistance is removed, i.e., $R = 0\ 1/\text{m} - \text{s}$. Compute the *total* solution for the fluid height, $h(t)$, as a function of time.

Problem 2 The height of fluid in a tank with two outlet pipes, one at the bottom of the tank and one 2 meters directly above it, is given by the following governing differential equation:

$$\frac{A}{g}\dot{h} + \frac{1}{R_1}h + \frac{1}{R_2}(h - H) = \frac{\dot{Q}_{IN}}{g}$$
$$A = 20\,\text{m}^2$$
$$H = 2\,\text{m}$$
$$R_1 = 2\ 1/\text{ms}$$
$$R_2 = 2\ 1/\text{ms}$$
$$\dot{Q}_{IN} = 30\,\text{m}^3/\text{s}$$
$$g \approx 10\,\text{m/s}^2$$

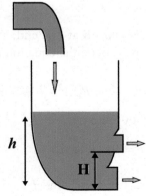

(a) What are the conserved quantity, and the effort and flow variables?

(b) Sketch the response for the height of fluid in the tank. Assume the initial height is 5 meters.

(c) Assume the system has already come to steady state. From this new initial state, what is the new steady-state height of fluid in the tank if the top outlet pipe is suddenly lowered 1 m, i.e., $H = 1$ m?

(d) How long will it take to attain this new steady-state height?

Problem 3 The differential equation governing heat transfer in the thermocouple probe quenched suddenly in a fluid bath maintained at T_∞ is given by

$$mc_P \frac{dT}{dt} = hA \ (T_\infty - T)$$

where m is the mass of thermocouple bead, c_p is its specific heat per unit mass, h is the heat transfer coefficient of still air, and A_S is the surface area of the thermocouple bead. Supposed it is known that the time constant for an experiment is 3 minutes from which it is determined that the heat transfer coefficient of still air is $h = 15\frac{W}{m^2-°C}$. If you know the heat transfer coefficient of still ice water is $h = 3600\frac{W}{m^2-°C}$, roughly how long will it take for the thermocouple bead to reach steady-state when the probe is re-immersed quickly into the ice water?

Problem 4 For the thermocouple probe quenched in Problem 3, the temperature, $T(t)$, is governed by the following 1st order ODE plunged suddenly in boiling water from standing air:

$$\frac{mc_P}{hA_S} \frac{dT}{dt} + T = 100$$
$$(T = 0) = 20\,°C$$

(a) Sketch the dynamic thermal response of this first order system of a room temperature mass suddenly placed in boiling water.

(b) Consider that at the same time, you have a second mass with double the heat capacity of the original mass, c_P, that is at an initial temperature of 150 degrees C when it is placed suddenly in a reservoir of liquid with a heat transfer coefficient 40% of that for water. On the same graph, sketch the response of this second mass.

(c) Write the functional form of the temperature solution for the second mass.

Problem 5 Consider an electrical analogy to a human artery provided by the 4-element Windkessel model. The capacitor represents the elasticity of the arterial wall, i.e., ranges of this value can model hardening of the arteries. The resistance to blood flow is determined by the viscosity of blood, i.e., a dehydrated patient will exhibit more viscous blood and a higher resistance to flow. An inductor is said to simulate inertia of the blood, i.e., it can model the density of blood changing as when its iron content becomes depleted.

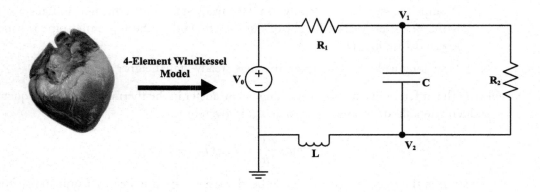

$$P(t) = 25 \cos \omega t \text{ V}$$
$$R_1 = 1000 \, \Omega$$
$$R_2 = 1000 \, \Omega$$
$$C = .002 \, \text{f}$$
$$L = 40 \, \text{H}$$

In this analogy, the current represents the blood flow rate, the applied voltage source represents the effort variable of blood pressure, and the frequency of the input excitation is the heart rate or pulse (where it is understood that rad/s correlates with beats-per-minute).

The governing differential equation for the system blood flow rate (current in the model) is given by:

$$LC \frac{d^2 i_C}{dt^2} + \left(R_1 C + \frac{L}{R_2} \right) \frac{di_C}{dt} + \left(1 + \frac{R_1}{R_2} \right) i_C = \frac{P(t)}{R_2} + C \frac{dP}{dt}$$

where $R_1 = R_2 = R$.

Consider that the so-called inertia of the fluid is small, but not zero. Mathematically, this implies

$$LC \ll RC$$

(a) Make a mathematically convincing argument, i.e., back it up with the necessary equations/relationships, to show that as the heart rate increases dramatically, a condition known as tachycardia, the blood flow rate decreases for a given constant blood pressure. Assume any response to initial conditions has decayed away and the system is in steady state.

HINT: Consider the transfer function for $I/(P/R)$ when formulating your answer!

(b) Describe, in words, the behavior of the amplification ratio, $I/(P/R)$, at low ($r << 1$), intermediate ($r \approx 1$), and high ($r >> 1$) normalized frequencies where $r = \omega/\omega_N$.

(c) For the given input magnitude blood pressure of 25 V, if a life-viable cutoff blood flow rate in steady state is 4 milli-amperes, at what heart rate, ω, will the patient expire?

Problem 6 Consider an older weightlifter who loves sausage and whose diet has hardened his arteries. An electrical 4-element Windkessel model identical in structural form to that for Problem 5 may be used. For such an analogous electrical heart, the governing differential equation and corresponding transfer function for the weightlifter's blood flow rate are given by:

$$\left(1 + \frac{R_1}{R_2}\right) I(t) + \frac{L}{R_2}\dot{I}(t) + LC\ddot{I}(t) = \frac{1}{R_2}P(t) + C\dot{P}(t)$$

$$\frac{I_{OUT}}{(P_{IN}/(R_1 + R_2))} = \frac{1 + \frac{r}{2\zeta}j}{(1 - r^2) + 2\zeta r j}$$

Assume: $L = 40\,\text{H}; C = 100\,\mu\text{f}; R_1 = 9000\,\Omega; R_2 = 1000\,\Omega; P(t) = 1000\cos(50t)\,\text{V}.$

(a) For these conditions, what is the steady-state amplitude of blood flow rate (current)?

(b) If the inductance is increased 5 fold to 200H and the capacitance is further reduced 5 fold to 20 μf (i.e., the arteries continue to harden), to what extent will this change the steady-state blood flow rate amplitude?

(c) Find an expression for the amplification ratio, $\mathcal{A}$, as a function of damping ratio, ζ, at resonance ($r = 1$).

Problem 7 Consider the electrical circuit analog for a quenched solid in an insulating jacket as shown here:

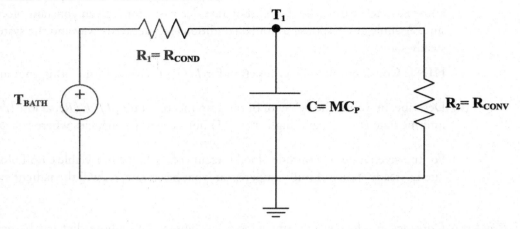

The resulting governing differential equation for the solid's temperature is given by:

$$\rho V c_P \frac{dT}{dt} + \left(\frac{1}{R_1} + \frac{1}{R_2} \right) T = \frac{1}{R_1} T_{BATH} \cos(\omega t)$$

(a) When forced by a periodic input at *low* frequency, the amplitude ratio $\dfrac{T}{T_\infty}$ approaches what value? Give your answer as an algebraic expression in terms of R_1 and R_2.

(b) What does the amplitude ratio $\dfrac{T}{T_\infty}$ approach for high frequency input?

Problem 8 Before insulating materials were readily available, buildings were thermally insulated by endowing their walls with sufficiently large thermal time constants.

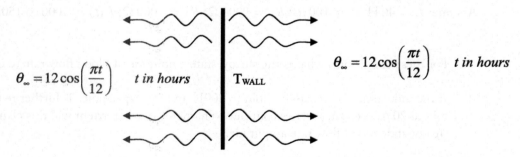

When driven by the daily solar thermal fluctuation, the differential equation governing θ, the fluctuation in wall temperature above and below its average daily value, is given by:

$$4\dot{\theta} + \theta = 12 \cos \left(\frac{\pi}{12} t \right) \,°\mathrm{F}$$

where the time, t, is measured in hours.

(a) What is the amplitude of the steady-state thermal fluctuation of the wall (in °F)?

(b) For what value of the thermal time constant will the amplitude of the steady-state thermal wall fluctuation drop to ±2°F, effectively insulating the building?

(c) How would you re-design the wall so that the steady-state response is reached in approximately 6 hours? You may state your answer in terms of the characteristic time or times of the system response.

(d) Draw an analogous electrical circuit whose behavior would be equivalent in some sense to this thermal problem. Label all the analogous electrical system elements corresponding to each of the thermal elements and describe the relevant input forcing function to the electrical circuit.

Problem 9 Consider Thomas Jefferson's home at Monticello, built before insulation was available. In the 18th century, buildings were thermally insulated by endowing their walls with sufficiently large thermal time constants.

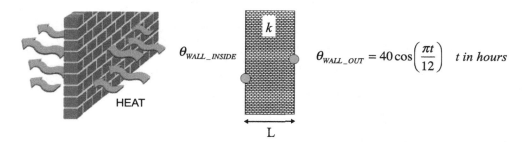

When driven by the daily solar thermal fluctuation, the differential equation governing θ, the fluctuation in wall temperature above and below its average daily value, is given by:

$$mc_p\dot{\theta} + \frac{kA}{L}\theta = \frac{40kA}{L}\cos\left(\frac{\pi}{12}t\right)\;°C$$

where the time, t, is measured in hours, and:

$$kA = 100\,\text{Jm/hr}\,°C$$
$$L = 0.25\,\text{m}$$
$$mc_p = 1600\,\text{J/}°C$$

The dimensionless Biot number, $Bi = \dfrac{hL}{k} = \dfrac{R_{COND}}{R_{CONV}}$ quantifies the relative magnitudes of conductive and convective resistances in a thermal system. The wall here is designed such that its Biot number is very large so that convection to the air surrounding the walls can be neglected.

(a) What is the amplitude of the steady-state thermal fluctuation of the wall (in °C)?

(b) For what value of the thermal time constant will the amplitude of the steady-state thermal wall fluctuation drop to ±5°C, effectively insulating the building?

(c) Re-design the wall by altering its thickness *only* so that the time constant obtained in part (b) can be obtained?

(d) With the time constant from part (c), how many hours will any transient now last en route to steady state?

(e) Draw an analogous electrical circuit whose behavior would be equivalent to this thermal problem. NOTE: You must draw the actual circuit and then label all the analogous electrical system elements corresponding to each of the thermal elements and describe the relevant input forcing function to the electrical circuit.

CHAPTER 8

Summary

The rules that describe nature seem to be mathematical. It is not a characteristic necessity of science that it be mathematical. It just turns out you can state mathematical laws which work to make powerful predictions. Why nature is mathematical is, again, a mystery.

Richard Feynman
The Meaning of It All

Fortunately, today's online world, with its advances in video and animation, offers several underused opportunities for the informal dissemination of mathematical ideas. Perhaps the most essential message to get across is that with math you can reach not just the sky or the stars or the edges of the universe, but timeless constellations of ideas that lie beyond.

Manil Suri
How to Fall in Love With Math

A la Suri [15], what we've sought to offer here is a digestible version of building governing differential equations from the cartoon building blocks of characters with whom are associated fundamental relations from the effort-flow analogy. We've presented an animated storyline wherein Captains Potential Energy and Kinetic Energy store system energy while the Evil Dr. Friction finds ways to steal it. We motivate these characters as roles in a common movie script about energy transfer in systems dynamics. We've then introduced the mechanical, electrical, fluid, and thermal casts that play these energy roles in the separate system disciplines.

It has been our intention to simply provide a mnemonic device to remember that separate physical actors always play the same roles in this movie. We also associate with these roles in the script equations relating effort and flow. Simple conservation balances then hopefully provide a more straightforward way to remember how to derive a governing differential equation for the system. We have also provided the story to show how features of our superheroes characterize the solutions to these equations. More than half of the students taught with these character representations of the effort-flow analogy claim these stories made coming to terms with systems dynamics more fun and the concepts more memorable. Learning can be fun. Even learning math can be fun!

Figure 8.1: The cast of the movie script for systems dynamics: Father Force, Captains Potential and Kinetic Energy, and the "not always Evil" Dr. Friction!

Afterword

This book has been written to present multi-disciplinary systems in a common light with an encompassing story focused on energy storage and dissipation. Based on our experience teaching the effort-flow analogy with these energy superheroes, we have found that the mnemonic of characters performing a common script played by discipline-specific actors helps students more clearly identify with the theme common to these dynamic systems. We have chosen a variety of chapter activities that illustrate this common behavior across engineering disciplines. After reading this manuscript, if you have comments on the presentation of the storyline or the orchestration of the chapter activities and examples or wish to suggest additional examples that emphasize system similitude across disciplines, feel free to contact the authors at COEcomments@gmail.com. Thank you, in advance, for any input you have.

Bibliography

[1] Dym, C. (2004). *Principles of Mathematical Modeling*. Academic Press. 16

[2] Feynman, R. P. (1998). *The Meaning of It All: Thoughts of a Citizen-Scientist*. Perseus Books. 1, 16

[3] Feynman., R., Gottlieb, A., and Leighton, R. (2006). *Tips on Physics:*. Pearson Addison Wesley. 1

[4] Feynman, R. (2009). Richard Feynman on Electricity. https://www.youtube.com/watch?v=kS25vitrZ6g. 22

[5] Feynman, R. (2012). What is the relationship between mathematics, science and nature? http://www.researchgate.net/post/What_is_the_relationship_between_Mathematics_Science_and_Nature. xiv, 1

[6] Jensen, B. D. and McLain, T. W. (2012). *System Dynamics*. http://twmclasses.groups.et.byu.net/lib/exe/fetch.php?media=483:335notes.pdf. xiii

[7] Johnson, A. T. (1998). *Biological Process Engineering: An Analogical Approach to Fluid Flow, Heat Transfer, and Mass Transfer Applied to Biological Systems*. Wiley-Interscience.

[8] Johnson, A. T. (2001). Teaching by analogy: The use of effort and flow variables. *Proceedings of the 2001 American Society of Engineering Education Annual Conference & Exposition*, Session 2973:1–3. xiii

[9] Lehrer, J. (2012). *IMAGINE: How Creativity Works*. Houghton Mifflin. xvii

[10] Ogata, K. (2003). *Systems Dynamics*. Prentice Hall. 73

[11] Palm, W. (2013). *Systems Dynamics*. McGraw Hill-Engineering-Math. 73

[12] Public Broadcasting System–NOVA (1993). The Best Mind Since Einstein - Richard Feynman Biography. Television Production. 16

[13] Singer, S. and Smith, K. A. (2013). Discipline-based education research: Understanding and improving learning in undergraduate science and engineering. *Journal of Engineering Education*, 00:1–4. DOI: 10.1002/sce.21091. xv

[14] Sofia, J. W. (1995). The fundamentals of thermal resistance measurement. Technical report, Analysis Tech. 16

[15] Suri, M. (2013). How to Fall in Love With Math. http://www.nytimes.com/2013/09/16/opinion/how-to-fall-in-love-with-math.html. 189

[16] Susskind, L. and Hrabovsky, G. (2013). *The Theoretical Minimum: What You Need to Know to Start Doing Physics*. Basic Books. 4

[17] Tippett, K. (2010). *Einstein's God*. Penguin Books. xiii

[18] Wellstead, P. E. (2000). *Introduction to Physical System Modelling*. www.control-systems-principles.co.uk. xiii

[19] Woods, R. L. and Lawrence, K. L. (1997). *Modeling and Simulation of Dynamic Systems*. Prentice Hall. xiii, 73

Authors' Biographies

VINCENT C. PRANTIL

Vincent C. Prantil earned his B.S., M.S., and Ph.D. in Mechanical Engineering from Cornell University where he was awarded the Sibley Prize in Mechanical Engineering and held an Andrew Dickson White Presidential Fellowship. He was a Senior Member of Technical Staff at Sandia National Laboratories California in the Applied Mechanics and Materials Modeling Directorates for eleven years. He joined the faculty in the Department of Mechanical Engineering at the Milwaukee School of Engineering in September 2000 where he presently specializes in finite element model development, numerical methods, and dynamic systems modeling. Since joining academia, he has become interested in the use of animation to both engage students and as a suggestive tool for students to use as a mnemonic device to enhance long-lasting learning. In addition to working with Tim Decker in Milwaukee, he has teamed up with colleagues at Northern Illinois University and Rutgers University in their efforts to showcase the power of video simulation for teaching undergraduate engineering concepts in dynamic modeling and controls theory.

TIMOTHY DECKER

Timothy Decker has played an important role in educational engagement over the past several decades. With extensive experience in game animation, character design and children's television, Tim has been an Animation Supervisor for Disney Interactive, lead animator for Knowledge Adventure, and layout artist/animator for the award-winning television series "The Simpsons" as well as Teenage Mutant Ninja Turtles, Alvin and Chipmunks, and the Critic. He has also appeared on many episodes of the "Imagination Station" as a guest artist inspiring children in the art of animation and cartooning. He has extensive experience directing animation in Canada, India, Korea, and the United States. Throughout his career, Tim has won numerous gaming awards from *PC Magazine*, *Communication Arts Magazine*, *Family Magazine* and the Academy of Arts and Sciences. Tim has been awarded three regional Emmy awards for his participation with Milwaukee Public Television. Tim holds a Bachelor's degree in Character Animation and Film from California Institute of the arts (CalArts) and an Associates degree in Illustration from Rocky Mountain College of Art and Design. Tim is enjoying his second career as a Lecturer at Peck School of the Arts at the University of Wisconsin–Milwaukee and Milwaukee Area Technical College. Tim teaches animation, character development, puppetry, claymation, and drawing for animation. His students are major participants in many national and international film festivals. Tim believes that immersive virtual environments are advantageous for communicating complex ideas, and that animation has the ability to support the telling of scientific stories in medical, engineering, and applied sciences.

CPSIA information can be obtained
at www.ICGtesting.com
Printed in the USA
LVOW05s0710050118
561939LV00005B/317/P